DESIGN OF HYDRAULIC GATES

Design of Hydraulic Gates

Paulo C.F. Erbisti
Consulting Engineer

A.A. BALKEMA PUBLISHERS / LISSE / ABINGDON / EXTON (PA) / TOKYO

Printed in the Netherlands by Krips The Print Force, Meppel, The Netherlands

Published by: A.A. Balkema Publishers, a member of Swets & Zeitlinger Publishers
www.balkema.nl and www.szp.swets.nl

ISBN 90 5809 621 1

CONTENTS

PREFACE   ix

ACKNOWLEDGEMENTS   xi

Chapter 1   INTRODUCTION   1

1.1   History and Development   1
1.2   Gate Components   10
1.3   Main Applications   11
1.4   Types and Classifications   12

Chapter 2   TYPES OF GATES   17

2.1   Flap Gate   17
2.2   Cylinder Gate   22
2.3   Stoplogs   23
2.4   Slide Gate   28
2.5   Caterpillar Gate   32
2.6   Miter Gate   35
2.7   Roller Gate   40
2.8   Segment Gate   42
2.9   Sector Gate   60
2.10   *Stoney* Gate   62
2.11   Drum Gate   64
2.12   Bear-Trap Gate   65
2.13   Fixed-Wheel Gate   67
2.14   Visor Gate   77

Chapter 3   BASIS FOR SELECTION OF GATE TYPE   79

3.1   Introduction   79
3.2   Most Common Types   79
3.3   Operational Requirements   80
3.4   Present Limits of Gate Sizes and Heads   81

Chapter 4   HYDROSTATICS   95

4.1   Introduction   95
4.2   Vertical Lift Gates   95
4.3   Radial Gates   102

Chapter 5   STRUCTURAL DESIGN   111

5.1   Load Cases   111
5.2   Allowable Stresses   113
5.3   Skin Plate   115
5.4   Horizontal Beams   120
5.5   Segment Gate   132

Chapter 6   EMBEDDED PARTS, GUIDES AND SUPPORTS   149

6.1   Slots and Niches   149
6.2   Wheel Track   152
6.3   Slide Track   160
6.4   Concrete Bearing Pressure   164
6.5   Lateral Guidance   169
6.6   Wheels and Pins   169
6.7   Bushings   180

Chapter 7   ESTIMATING GATE WEIGHTS   183

7.1   Introduction   183
7.2   Segment Gates   184
7.3   Fixed-Wheel Gates   185
7.4   Double-Leaf Fixed-Wheel Gates   186
7.5   Stoplogs   187
7.6   Flap Gates   188
7.7   Caterpillar Gates   189
7.8   Embedded Parts   190

Chapter 8   HYDRODYNAMIC FORCES   199

8.1   Introduction   199
8.2   Model Tests   202
8.3   Factors Influencing Downpull   205
8.4   Formulae for the Prediction of Downpull   208
8.5   Method of Knapp   219

Chapter 9   GATE OPERATING FORCES   225

9.1   Introduction   225
9.2   Gate Weight   225
9.3   Friction on Supports and Hinges   226
9.4   Seal Friction   228

Chapter 10   AERATION   235

10.1   Introduction   235
10.2   Air Vent – Functions and Features   236

10.3   Air Vents – Empirical Calculation   237
10.4   Air-Demand Ratio   238
10.5   Air Vent Dimensioning   243

Chapter 11   GATE HOISTS   249

11.1   Introduction   249
11.2   Screw Lifts   249
11.3   Wire Ropes   251
11.4   Roller Chains   255
11.5   Oil Hydraulic Drives   258
11.6   Gate Hoist Arrangement   265
11.7   Hand Operation   268
11.8   Design Criteria   268
11.9   Gate Position Measurement   269

Chapter 12   MATERIALS   275

12.1   Introduction   275
12.2   Heat Treatment   275
12.3   Rolled Steels   276
12.4   Steels for Machine Elements   276
12.5   Stainless Steels   277
12.6   Cast Steels   277
12.7   Forged Steels   277
12.8   Gray Cast Irons   278
12.9   Bronzes   278
12.10   Bolts   278

Chapter 13   GATE SEALS   285

13.1   Introduction   285
13.2   Wood Seals   285
13.3   Metallic Seals   285
13.4   Rubber Seals   285
13.5   Material for Rubber Seals   289
13.6   Clad Seals   289
13.7   Rubber Seal Hardness   290
13.8   Rubber Specifications   290
13.9   Seal Leakage   291
13.10   Manufacture and Assembly of Seals   291

Chapter 14   MANUFACTURE, TRANSPORTATION AND ERECTION   305

14.1   Manufacture   305
14.2   Transportation   317
14.3   Field Erection   318
14.4   Acceptance Tests322

Chapter 15   TRENDS AND INNOVATION IN GATE DESIGN   325

15.1   Long-Span Gates   325
15.2   High-Head Gates   330
15.3   Refurbishment and Modernization of Gates and Dams   332

NAME INDEX   343

SUBJECT INDEX   345

PREFACE

Nobody is better qualified than Prof. Paulo Erbisti to have written this book "Design of Hydraulic Gates", because he has acquired a very broad personal experience during his career devoted to hydroelectric projects, not only with a scientific spirit, but also a practical approach. He knows very well all the phases of implementation of hydromechanical equipment, from feasibility studies through to erection and commissioning, as well as rehabilitation.

It is a great advantage for engineers throughout the world that the second edition of this book has been produced in English. But we can understand very well the value of the first edition published in Portuguese in 1987, because the extent of experience in Brazil in implementing hydraulic works, particularly hydro plants, is considerable (and Prof. Erbisti has worked in various positions of responsibility for major electricity authorities in his country). It should be recalled that Brazil has about 64,000 MW of hydro capacity in operation today, supplying about 300 TWh/year, or around 94 per cent of national electricity production; a further 25,000 MW of additional hydro capacity is currently planned. So, the specific experience of Brazil is in fact a reference for the world.

This book is extremely welcome as an efficient synthesis of a broad subject, and engineers from the various disciplines, not only mechanical and electrical but also civil engineers, will recognize its value. In western countries, the civil works of a hydropower scheme typically represent 70 to 80 per cent of the costs, and in this respect the electro-mechanical works do not rank in the first place of concern. But experienced engineers know well the importance of good design, construction and operation of equipment, and particularly gates, which are often directly connected with the safety of the civil works. Over the past 20 years, the majority of dam failures caused by overtopping resulted from the malfunctioning of spillway gates.

The author has covered the subject of gates very comprehensively in his book, from history to new developments. He discusses not only the principle of gates and their associated equipment, the choice of type and calculation methods, but also he (as an excellent practical engineer, who knows that 'the devil hides in the details') describes their construction details. Each chapter is supported by a complete list of international references.

The book will be very useful for experienced engineers (as well as students) for the conception, construction and operation of gates. Gateworks have had a long historical development, and they have a great future. Only about 20 per cent of the economically feasible hydro potential of the world has been exploited. The present installed hydro capacity of around 750 GW will be multiplied by a factor of 3 during this century. The number of large dams (45,000 today) necessary for water supply and irrigation as well as hydropower, will need to be doubled.

The interest of this book is evident as a reference, and Prof Erbisti should be gratefully acknowledged for his important contribution to this technology.

Raymond Lafitte
Professor at EPFL (Federal Institute of Technology, Lausanne)
President of the International Hydropower Association

ACKNOWLEDGEMENTS

This is the English version of the second edition of the book *Comportas Hidráulicas* published in Portuguese in 2002 in Brazil. It reflects my experience as a design engineer and a consultant in a country where more than 90% of the dams responsible for the current hydroelectrical generating capacity have been designed and built over the last three decades. This rapid expansion in energy production transformed Brazil into a country with a diversified industrial economy and an installed electrical capacity similar to that of the United Kingdom.

My goal with this book is to gather into a single document all the necessary information to select and design a gate. I have tried to present simple answers to the following questions: How and when did the first gates develop? What are they? How do they work? How do you choose the most adequate type? How do you design the gate structure and the support elements? How much force is necessary to move the gate? How do you manufacture them? What are the precautions to be taken during gate erection? What are the new developments and the latest trends in gate design and manufacture?

To illustrate the text, a large number of drawings and photographs of equipment have been included. Long lists of manufactured gates are presented with indications of their main characteristics such as: dimensions, acting hydrostatic load, weight, supply date, and name of the supplier. The structural and mechanical calculation methods follow the recommendations of the main Brazilian and German standards for equipment design. For the structural design, conservative factors of safety are recommended for the admissible stresses, which will result in robust and reliable equipment. Based on the analysis of statistical data, equations were developed to estimate the gate weights taken from their basic dimensions. The main constructive aspects of exceptionally large or very deep gates are analysed; various ways to heighten gates and dams are shown, including the use of the most recent technological advances such as inflatable gates and fusegates.

I would like to thank the companies and institutions which kindly allowed the use of their papers, tables and experimental coefficients, with emphasis to: ASCE, ASHRAE, U.S. Army Corps of Engineers, ICOLD, SOGREAH, La Houille Blanche and G. Braun Verlag. I am particularly grateful to engineer John Cadman for his valuable contribution on the revision of the English text, and to engineer Marlos Fabiano de Moraes for his great help in formatting the illustrations included in this book. Finally, I extend my sincere thanks to all the manufacturing companies who provided the photographs, drawings and catalogues of their products for this book: Bardella, Alstom, Dedini, Voith, Rubberart, Rexroth Hydraudyne, H. Fontaine, Rittmeyer, Glacier, Hydroplus, VA TECH Hydro, Rodney Hunt, DSD-Nöell, Giovanola Frères, Hitachi-Zosen, Mitsubishi, Sumitomo, MAN, Krupp, Thyssen-Klönne, IHI and ZWAG.

Paulo C. F. Erbisti
Rio de Janeiro, Brazil

Chapter 1

Introduction

1.1 HISTORY AND DEVELOPMENT

The construction of hydraulic gates was closely related with the development of irrigation, water supply and river navigation systems. In the early days of hydraulic engineering, water was backed up by small dams and conveyed to side irrigation canals. The excess water was discharged over the dam. As a natural evolution, 'movable dams' were built. These movable dams could be removed from their normal position to provide passage for excess water, thus permitting greater safety and flexibility in the operation of hydraulic works.

The first canals for transportation of goods and drainage of floodwaters were built in China. Originally, the Chinese solved the problem of fluvial transportation in the region of river rapids by building dikes with slopes on the banks of the canal. The boats were then manually hoisted up and down the slope. These operations, however, were both time- and power consuming. Around the year of 983, the Chinese discovered that by constructing two dams a certain distance apart, the boats could enter the 'pool' created between them and the water level could be slowly increased or decreased. The earliest dams had wood or stone piers on each side of the canal. Vertical grooves were cut into opposite sides of the banks and tree trunks were fitted horizontally into the grooves, which held the water at the highest level. Ropes were used to lift the trunks. Later, the trunks were linked, forming an integral barrier that could be lifted or lowered like a guillotine blade.

The development of gates in the Netherlands followed a pattern similar to that of China. At the end of the 14th century, locks were very common there. The gates, still of the guillotine type, were provided with lead counterweights and equipped with drains, which permitted emptying gradually the lock chambers [1].

In 1795, the Little Falls canal was completed, making it the first canal with locks in America. Design of wooden gates for the Little Falls locks was unusual. Two wood swinging gates were placed at each end of a lock. Instead of closing to a flat plane, the gates closed to form an angle pointing upstream, facing the current. Water pressure thus locked them together. Near the base of the large gates were sluice jacks or small cast iron plates, which pivoted vertically, allowing water to enter or leave a lock. The same arrangement, using miter gates, was used for the Great Falls locks. Butterfly valves, pivoting horizontally at mid-height were used instead of sluice jacks to empty and fill the lock chambers [2].

The first metal gates appeared around 1830. With the turn of the century, various inventions occurred as well as a great development of the existing types, furthered by the challenge of the need to build ever-larger gates.

Filipo Maria Visconti designed the first pound lock in 1439. This was at Vareno, near Milan, Italy, to improve navigation for the transportation of granite blocks used in the construction of the Milan's Duomo. In 1638, a pound lock was built at the Briare canal in France [3]. The first drawing of a pound lock is dated 1497 (see Figure 1.1). This illustration already exhibits the main features of a modern lock, with gates pivoted instead of working vertically. The enlargement in the center permitted the passage of more boats at the same time.

Fig. 1.1 First illustration of a lock (1497)

On the chamber walls anchorage eyes were installed to which the crafts could be connected by ropes to prevent their displacement during the chamber filling and emptying, operations that certainly caused great turbulence. One of the lock gates is a miter gate similar to those used nowadays. Filling and emptying were carried out through small openings provided with gates for their closure. This system is still used at the present time in small locks [4]. Miter gates, designed with cast iron structure and steel plate shielding, were used as far back as 1828, on the Nivernais canal in France. A metallic miter gate was used in the Charenton lock, France, in 1864; this gate was 7.8 m wide by 7.76 m high [5].

The invention of the pound lock is credited to Leonardo da Vinci (1452-1519). This is not true, although da Vinci did bring many innovations to it.

The oldest known application of a segment gate was in 1853, on the Seine River, in Paris, where four gates 8.75 m wide by 1.0 m high were installed. These were designed by the French Engineer Poirée [6], who is also the inventor of the needle dam first used in 1834 on the Yonne River, in France [7]. Other early applications occurred on the Nile River delta, near 1860, where 132 segment gates 6 m wide by 5.1 m high were built by the French Engineer Mougel Bey for the Rosetta and Damietta dams. The gate arms were subjected to traction. At the time they were called *'cylindrical gates with radiuses subjected to traction'*. According to Wegmann, the original gates for closing the openings between the piers were shaped as the arc of a circle, supported at either end by iron rods radiating from the center of the arc, where they were attached to massive iron collars, working round cast-iron pivots embedded in the masonry of the piers. These gates were to be lowered by their own weight and to be raised by compressed air pumped into the hollow ribs, but they could not be operated successfully and were replaced after 1884 by wrought-iron gates provided with rollers sliding in cast-iron grooves fixed in the piers, according to F. M. Stoney's patent. Powerful crab winches (two for each dam) traveling on continuous rails served to lower or raise the gates [8]. In 1910 a patent of a reverse segment gate was assigned in the USA to L. F. Harza.

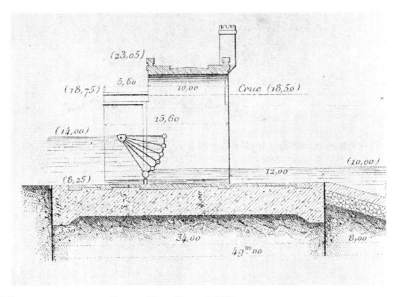

Fig. 1.2 Reverse segment gate, Rosetta Dam, Egypt (1860)

Around 1870 in the USA parallel inventions of the segment gate occurred [9]. Rehbock and Hilgard, together with A. O. Powell, give the name of the inventor as T. Parker who, however, presumably sold his ideas to Jeremiah Burnham Tainter, from Menomonee, Wisconsin. In 1886 he patented it in his name, receiving from the U. S. Patent Office the number 344879. The gate had three radial arms and wood construction. It was driven by chains installed upstream of the skin plate. The gates would be installed in tandem to serve as lock gates and also for filling and emptying the lock chamber.

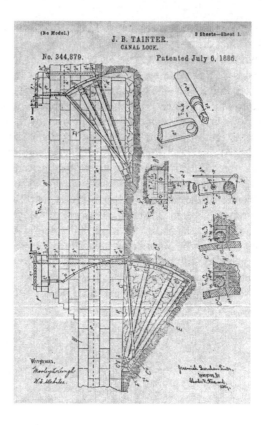

Fig. 1.3 Segment gate, Tainter's patent (1886)

Figure 1.4 shows an interesting application of a segment gate with counterweights, built in 1888, at the Lez River, south of France.

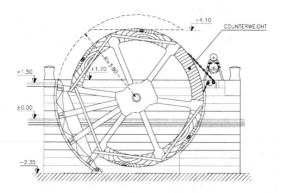

Fig. 1.4 Segment gate with counterweights, Lez River, France (1888)

In Germany, the first reports of segment gates date from 1894/1895 and refer to the installation of a gate with a span of 12 m and 1.87 m high in the Werderschen Mühlegraben, Berlin. Later on, in 1903, another gate was built on the Landwehr canal, with a 5.56 m span and 1.6 m high. Both had arms subjected to compression. In 1895, reports were published in the USA describing the use of segment ('*Tainter*') gates in the Illinois-Mississippi channel.

The segment gates were initially used in the USA for flow control in conduits and used for the first time in lock aqueducts in the construction of the New York barge canal, in 1905. Reverse segment gates were used again in 1953, at the Oberpeichning dam, on the Lech River, Germany. They were 16 m wide and 8.25 m high, and had a 2.15 m high flap gate at the top. Because of the highly reliable performance of these gates, about 28 others were built up to 1976 in the Bavaria region, Germany [4].

Notwithstanding the efforts developed in the 19th century by the various inventors of the segment gate, it is remarkable that all them were preceded by the great genius of the Renaissance, Leonardo da Vinci. In his studies on hydraulics, around 1490, da Vinci already registered that type of gate. In Figure 1.5, one of his studies in Codex Atlantis, it can be found: miter gate (right); top-hinged flap gate (below, right); plain gate with vertical hinges (left); and a segment gate (top, center). This gate has radial arms extending beyond the trunnions. The arms act as counterweights, easing the manual operation of the gate by leverage. The gate shield is curved like the modern gates.

Fig. 1.5 Various types of gates, by Leonardo da Vinci (circa 1490). Property of the Ambrosian Library. All rights reserved. Reproduction is forbidden

The much-emphasized advantage of the segment gate (the absence of slots in the piers) does not appear in the first reports because then they were often used. In 1914, H. Engels said: *'The segment gates present advantages over other gate types up to 12 m of span'* [10]. H. Kulka, in 1928, was even more optimistic: *'The introduction of forces in the piers does not cause problems, even on large gates. There is no problem in the construction either of the trunnions or of the piers. Segment gates may be used for any practical dimension'* [11]. In fact, some segment gates have been built with remarkable dimensions:

Dam	River	Year	Span (m)	Height (m)
Barthelm	Oder	1920	40	3.0
Ladenburg	Neckar	1927	36	5.5
Münster	Neckar	1927	23	7.4
Donzère-Mondragon	Rhone	1948	45	9.0
Haringvliet	Rhine-Mosel-Scheld delta	1967	56.5	10.5
Vilyu		1967	40	14.0
Stör	Stör	1975	43	13.0
Altenwört	Danube	1976	24	15.5
Itaipu	Paraná	1982	20	21.3

Double-leaf gates originated in Europe, and are found in Japan. Double-leaf metal gates, 5 m wide by 5 m high, were used to close the 111 arches of the Assiout dam, Egypt, in 1902 [12].

In 1908, ten double fixed-wheel gates were installed on the Augst-Wyhlen dam, on the Rhine River, Switzerland, each gate being 17.5 m wide and 9 m high [13]. The modern double fixed-wheel gate of the hook type was developed by M.A.N. and installed for the first time on the Reckingen dam, Switzerland, in 1930.

Similar to the double-leaf fixed-wheel hook type gate, double segment gates were developed. However, few installations were built, all in Switzerland. These are the three of the Rupperswill-Auenstein dam, on the Aare River, in 1943, with a 22 m span and 8 m high and the two at the Brunau dam, on the Sihl River, in 1969, with a 20 m span and 5.5 m height [14].

The sector gate was invented in the USA by C. L. Cooley, and used for the first time in 1907 in the Lockport dam on the Chicago drainage canal. Two gates were installed to regulate the flow in the canal and to carry off ice and floating debris. One had a 3.66 m span, and the other a 14.6 m span. In both gates the curvature radius of the skin plate was 7.92 m and the height 5.79 m. The pivot consisted of lengths of 100 mm rods, supported by cast steel brackets on the back wall [8].

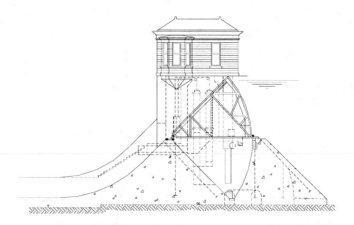

Fig. 1.6 Sector gate, Lockport dam (1907)

In Europe, the first application of the sector gate was in 1911, when two gates were installed on the Weser dam, near Hemelingen, Germany, with a 54 m span and 4.6 m height. According to H. Ackermann, during the early studies of that dam, the use of rolling, fixed-wheel, segment and flap gates was also studied. The choice of the type, however, favored the sector gate because of the need for a submersible gate. At that time, submersible gates of the roller, fixed-wheel and segment types were already known, but had not yet been tested in practice. The design of the Weser dam gates is similar to that of Lockport [15]. In 1924 three sector gates were installed in Brazil at the Ilha dos Pombos Dam, on the Paraiba River, with a 45 m span and a height 7.4 m. These gates so far hold the record of the largest impounding area (333 m²) for those of its type [16].

Max Carstanjen, chief engineer of the Gustavsburg Bridge Works (M.A.N.), Germany, invented the roller gate in 1898. Its first application was on the Sau River. The gate had an 18 m span and a height of 4.14 m. To reduce the buoyancy effect, its central body had a pear-shaped cross-section, while the ends were cylindrical. In 1903, another gate, 35 m wide by 2 m high, was installed in the same region, on the diversion canal of the Main River; this operated for about 60 years [17].

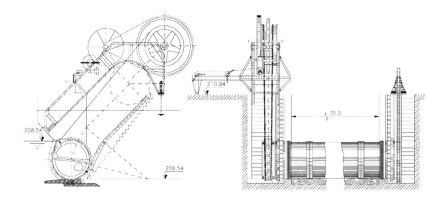

Fig. 1.7 Roller gate, Schweinfurt dam (1902)

The original roller gates were plain cylinders. Experience with the operation of such gates installed at Poppenweiler, on the Neckar River, Germany, indicated the occurrence of excessive vibrations as a result of the suction effect under the lifted cylinder. Later, a section of curved skin plate in the shape of an apron was added to the lower portion of the cylinder, assuring the separation of the underflowing nappe. The high torsion resistance inherent in the cylinder shape allowed for the design of gates consisting of a slightly curved shield placed upstream of a small cylinder [18]. Roller gates were often used in the USA, designed by the U. S. Corps of Engineers, mainly on the Ohio and Mississippi rivers, and by the Bureau of Reclamation. One of the earliest applications was in 1915 on the Colorado River, near Grand Junction, Colorado, where gates 21.3 m wide by 3 m high were installed [19]. For the canalization of the Mississippi, the U. S. Corps of Engineers installed roller gates with 18.3 m, 24.4 m and 30.5 m spans. In 1928 the Engineering News Record magazine recorded the installation of about 250 roller gates, in the following countries [20]:

Germany	140
Finland	20
Norway	19
Sweden	16
Czechoslovakia	13
Rest of Europe	14
USA	24
Mexico	4
South America	3

H. M. Chittenden, of the U. S. Army Corps of Engineers, invented the drum gate in the USA, in 1896. Its first application was at Dam No. 1 on the Osage River, USA, in 1911 [8]. The interior framework of the gate was made of iron, while the outside consisted of wood. The lower face and the ends of the gate were closed.

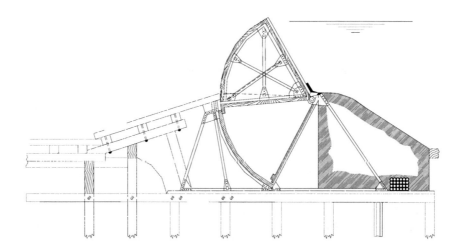

Fig. 1.8 Chittenden drum gate (1911)

In 1818, Josiah White constructed the first bear-trap gate in Mauch Chunk Creek, on the Lehigh River, USA, to increase the depth of water and improve the navigation of coal-transporting boats. The gate attracted much attention during its construction. The workmen at the site tried to rid themselves of curious people, who wished to find out what was being constructed, by telling them that it was a *'bear-trap'*, a name the gate has borne ever since. In 1819 twelve dams with gates of this kind were constructed on the Lehigh River, and these proved to be perfectly successful [8]. Bear-trap gates were also built in France, Germany and Switzerland.

The ring gate is a variation of the cylinder type and was developed by the Bureau of Reclamation, USA, for use in morning-glory spillways [21].

Only two applications are known, both in the USA: Owyhee dam (in 1936), with a diameter of 18 m and a height of 3.6 m; and Hungry Horse dam (in 1953), with a diameter of 19.5 m and a height of 3.6 m.

The *Stoney* gate is named after its inventor, Mr. F. G. M.Stoney. These gates were first constructed in 1883 at Belleek, Ireland. Four sluice gates were placed between masonry piers to control the outfall from a chain of lakes. Each gate was 8.9 m wide and 4.4 m high. They were made of plate iron and beams. *Stoney* gates were extensively used in Europe, USA, Egypt and India at the beginning of the 20th century. The only use of *Stoney* gates in Brazil was in 1925, at the Ilha dos Pombos dam. Three gates were installed at the intake structure and eight in the spillway. All these gates are still functioning.

Table 1.1 was based on the specialized bibliography and on reference lists of the main world gate manufacturers.

Table 1.1 Inventions and Earlier Use of Gates

Year	Gate type	Project	Span x height (m)	Inventor and/ or supplier
1490	Segment			L.da Vinci (?)
1818	Bear-trap	Mauch Chunk Creek	7.6 (span)	Josiah White
1828	Miter (metallic structure)	Nivernais, France		
1853	Segment	Senne River	8.75 x 1.00	Poirée
1860	Reverse segment	Nile River Delta	6.00 x 5.10	Mougel Bey
1873	Bottom-hinged flap	Ile Brûlée, France (*)	3.52 x 1.97	Girard
1883	*Stoney*	Belleek, Ireland	8.90 x 4.40	F. Stoney
1886	Segment	(USA patent)		J. B. Tainter
1896	Drum	(USA)		H.Chittenden
1898	Roller	(Germany)		Carstanjen
1902	Double-leaf fixed-wheel	Assiout, Egypt (*)	5.0 x 5.0	
1902	Roller	Schweinfurt	18.00 x 4.14	M.A.N.
1907	Sector	Lockport Dam, USA	14.60 x 5.79	C. L. Cooley
1910	Reverse segment	(USA patent)		L. F. Harza
1911	Drum	Dam no. 1, Osage River		H.Chittenden
1911	Sector	Weser Dam (*)	54 x 4.6	
1915	Broome (caterpillar)	Turner Falls, USA (*)	3.00 x 4.88	
1920	Cylinder (*)	Kern	2.43 dia. x 6.70	
1926	Fixed-wheel with flap	Juliana canal (*)	23.00 x 4.40	T. Klönne
1930	Double-leaf hook type	Reckingen		M.A.N.
1933	Segment with flap	Münster (*)	23.30 x 7.40	M.A.N.
1936	Ring	Owyhee Dam	18.0 dia. x 3.60	BuRec
1943	Double segment	Rupperswill-Auenstein	22.00 x 8.00	C. Zschokke
1955	Inflatable	(France)		Mesnager
1958	Reverse segment with flap	Oberelchingen (*)	16.00 x 8.20	M.A.N.
1960	Visor	Hagenstein, Holland	48.00 x 6.00	
1966	Inflatable	Shiga, Japan (*)	0.45 x 3.7	Sumitomo
1991	Fusegate	Lussas Dam, France	3.5 x 2.15	Hydroplus

(*) Refers to earlier use, but not necessarily to the first application.

1.2 GATE COMPONENTS

A gate consists basically of three elements: leaf, embedded parts and operating device. The leaf is a movable element that serves as bulkhead to the water passage and consists of skin plate and girders. The shield plate directly responsible for the water dam is called the skin plate. The seals, the components responsible for the water tightness, consist generally of rubber strips screwed on to the skin plate. On the gate leaf are also attached the support elements (wheels, rollers, bearing plates and so one) and guides (shoes, wheels, springs etc).

The embedded parts are the components embedded onto the concrete, which serve to guide and house the leaf, to redistribute to the concrete the forces acting on the gate, acting also as protection to the concrete edges and support element for the seal. The basic components of the embedded parts are (see Figure 1.9): sill beam, wheel or slide tracks, side guides, counterguides, lintel, seal seats and, eventually, slot lining.

The sill beam is the lower horizontal element of the embedded parts and serves as support for the gate leaf or the bottom seal. In sector, drum and bear-trap gates, the sill beam usually serves also as fixation of the bottom seal. The wheel track acts as support element and distributor of the loads transmitted by the wheels or rollers.

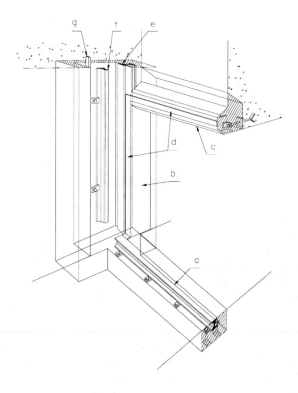

Fig. 1.9 Embedded parts components
(a) sill; (b) slot lining; (c) lintel; (d) seal seat; (e) wheel track; (f) side guide; (g) counterguide

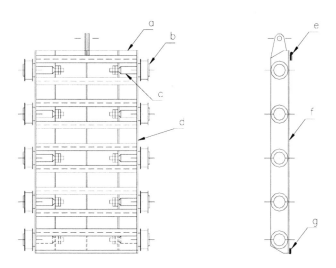

Fig. 1.10 Fixed-wheel gate nomenclature
(a) gate leaf; (b) wheel; (c) wheel pin; (d) end girder; (e) top seal; (f) skin plate; (g) bottom seal

The same function is exerted by the slide track in respect to the loads transmitted by the end vertical girders of stoplogs and slide gates. The side guides and counterguides limit the displacement of the gate leaf on the horizontal plan and are designed to absorb the corresponding stresses. The lintel is an element used only in submerged gates and serves to complete, along with the side guides and the sill, the water passage frame. It is located on the upper horizontal portion of the passage to be closed, supports the upper seal and serves also for the protection of the concrete against erosion caused by the water flowing at high speed.

The operating device is the means directly responsible for the opening and closure of the gate. Some gates dispense with the use of operating hoists and are moved by water pressure, such as sector, drum and bear-trap gates.

1.3 MAIN APPLICATIONS

Gates find a wide range of application in the various fields of hydraulic engineering. Among their main uses one may list:
a) flood protection works;
b) protection of equipment - emergency gates installed upstream of turbines;
c) level control in reservoirs intended for recreation or located near residential or non-flooding areas;
d) maintenance of constant level in reservoirs;
e) cleaning of reservoirs - discharge gates for floating debris (tree boughs, trunks and ice);
f) flow regulation in dams;

g) equipment maintenance - stoplogs installed downstream of turbines, upstream of spillway gates etc;

h) closure of river diversion conduits;

i) intake works for hydroelectric plants, water supply etc;

j) irrigation - water intake, head control, distribution etc;

k) control of bottom discharges;

l) navigation dams - lock gates, filling and emptying systems etc.

1.4 TYPES AND CLASSIFICATIONS

The terms usually adopted for the various gate types are: flap, cylinder, stoplog, slide, caterpillar, miter, roller, segment, sector, *Stoney*, drum, bear-trap, fixed-wheel and visor gates. Other gate types have been developed, especially in the first half of the century, but will not be discussed here for their use is rather rare. According to their features, the gates may be grouped of various manners. Among others, the following classification criteria may be listed: purpose, movement, water passage, leaf composition, location and skin plate shape.

1.4.1 PURPOSE

According to their operational purpose, gates may be classified as:
- service gates;
- emergency gates;
- maintenance gates.

Service gates are used for continuous regulation of flow or water level. Examples:
. spillway gates;
. bottom outlet gates;
. lock gates (navigation chambers and aqueducts);
. flood control automatic gates.

The emergency gates are used occasionally to shut down the flow of water in conduits or canals; as a rule they are designed for normal operation in open or closed position. Only in special situations can these gates be used in partial openings as, for instance, the intake gates where the filling of the penstock is provided through slight lifting of the leaf (operation called 'cracking'). Among others, the following are considered emergency gates:
. intake gates;
. gates installed upstream of penstock service valves;
. draft tube gates of Kaplan turbines;
. gates installed upstream of bottom outlet gates.

Maintenance gates are operated only under balanced pressure of water and their main function is to allow the emptying of the conduit or canal for access to and maintenance of the main equipment (turbine, pump, or even another gate). The most common type of maintenance gate is the stoplog.

1.4.2 MOVEMENT

According to the movement of the gate along its guides, the gates are classified as:

a) translation gates:
- sliding: slide, stoplog, cylinder;
- rolling: fixed-wheel, caterpillar, *Stoney*.

b) rotation gates: flap, miter, segment, sector, drum, bear-trap, visor;
c) translo-rotation gates: roller.

Translation gates may be either sliding or rolling. In the sliding gates, the gate leaf moves along the guides overcoming the friction of sliding between embedded and movable parts through shoes or bearing plates, whereas the rolling gates use wheels or rollers for that purpose.

In the rotation gates, the gate leaf turns around a fixed axis, called hinge axis. In the flap, sector, drum and bear-trap gates, the hinge axis lies on the sill, in a horizontal position. In the miter gates, the hinge axes (one for each gate leaf) are vertical and located near the lock chamber walls. Visor and segment gates are provided with a horizontal hinge axis located above the sill. In some rare cases segment gates were designed with vertical hinge axes assembled in pairs to serve as lock gates, as in the La Rance lock, in France.

The roller gate is the only gate that performs a combined motion of rotation and translation. Its leaf is a cylindrical structure with a horizontal axis that turns in a rack gear installed in an inclined recess in each end pier.

1.4.3 WATER PASSAGE

According to the water passage in relation to the leaf position, the following situations may occur:
a) discharge over the leaf - flap, sector, bear-trap and drum gates, in the opening operation they move down around the hinge axis located on the sill, permitting the water passage over the gate;
b) discharge under the leaf - slide, caterpillar, roller, segment, fixed-wheel, visor and *Stoney* gates move upwards, making possible the flow of water under the gate;
c) discharge over and under the leaf - mixed and double gates permit discharge alternately over and under the leaf, according to the operational requirements.

1.4.4 GATE LEAF COMPOSITION

Gates may be plain, mixed or double, depending on the amount and type of elements that comprise the leaf. The plain gates have the leaf with only one element. In the mixed ones, the main leaf has, at the top, a flap gate. Many applications are known of segment, fixed-wheel, roller and *Stoney* combined with flap gates, mainly in Europe. On double gates the leaf comprises two movable overlapping elements. The lowering of the upper element permits discharge over the gate, while the lower element can be lifted to discharge as an orifice. Both elements are raised for passage of the maximum flow. Fixed-wheel and segment gates are the only known types of double-leaf gates.

1.4.5 LOCATION

Depending on the location of the opening with respect to the headwater, gates may be either of weir or submerged-type gates. Submerged gates are necessarily provided with sealing all over the water passage perimeter, while weir gates do not have sealing on the upper edge.

All types of gates can be used on crest works, but only a few of them are applied on submerged installations, namely: fixed-wheel, segment, caterpillar, slide, stoplog, cylinder and *Stoney*. According to the water head over the sill, gates are usually classified as:
- low head gates: up to 15 m;
- medium head gates: from 15 m to 30 m;
- high head gates: over 30 m.

However, the criteria based on the water head are subjective and change according to the technological evolution.

1.4.6 SKIN PLATE SHAPE

Gates may be flat or radial, according to the shape of the skin plate. The following types are designed with flat skin plate: slide, caterpillar, fixed-wheel, *Stoney*, stoplog and bear-trap gates. The radial gates are: segment, sector, drum, visor, cylinder and roller gates. Flap and miter gates may have a flat or curved skin plate. Reverse segment gates, very common in Germany, most times have a flat skin plate.

REFERENCES

1. *Short History of the Inventions* (in Portuguese), Editora Abril Cultural (1976), São Paulo.
2. Great Falls Canal and Locks: Civil Engineering Land Mark, *Civil Engineering*, ASCE (Nov. 1972).
3. Turazza, G.: *Costruzioni Idrauliche*, Casa Editrice Dottor Francesco Vallardi (1900), Milan.
4. Roehle, W.: Historische Entwicklung Aufgabenstellung und dere Lösung im Bau von Schleuse in Wasserstrassen, *Stahlbau und Rundschau*, Heft 29 (1966).
5. de Mas, F.B.: *Rivières Canalisées*, Editeur Béranger (1903), Paris.

6. Csallner, K.: *Strömungstechnische und Konstruktive für die Wahl Zwischen Druck und Zugsegment als Wehrverschluss*, Versuchsanstalt für Wasserbau der T. U. München, Bericht Nr. 37 (1978), Obernacht.

7. Streiffer, A.: Tainter Gates of Record Size Installed in Spanish Dam, *Civil Engineer* (March 1950).

8. Wegmann, E.: *The Design and Construction of Dams*, John Wiley and Sons, 6th edition (1911).

9. Hartung, F.: Neuezeitliche Gesichtspunkte im Grosswehrbau, *Zeitschrift der Vereinigg. Deutscher Elektrizitätsw.* – VDEW, Band 59, Heft 15 (1960).

10. Engels, H.: *Handbuch des Wasserbaues*, 1 Band 1/3 Aufl., Vlg. v. Wilhelm Engelman (1914/1923), Leipzig.

11. Kulka, H.: *Der Eisenwasserbau*, Verlag v. Wilhelm, Ernst und Sohn, (1928) Berlin.

12. Barois, J.: *Les Irrigations en Égypte*, Editeur Béranger (1904), Paris.

13. Kollbrunner, C.F.: *Hydraulic Steel Gates*, Lehman Publishers (Sept. 1950), Zurich.

14. Streulli, L.J.: Movable Double-Segment Gates for Ice Retaining on the Sihl, near Zürich (Switzerland*), Acier-Stahl-Steel*, No. 4 (1976) pp.130-136.

15. Ackermann, H.: Vom Werden des Sektor-Wehres in Deutschland, *Die Bautechnick*, 38 Jahrgang, Heft 5 (1961), pp.145-149, Berlin.

16. Erbisti, P.C.F.: Segment or Sector Gate? (in Portuguese), *Construção Pesada* (Nov. 1979), São Paulo.

17. Schleicher, F.: *Manual del Ingeniero Constructor*, vol. II, Editorial Labor (1948), Madrid.

18. Schoklitsch, A.: *Construcciones Hidraulicas*, Editorial Gustavo Gili S.A., 3rd edition (1968), Barcelona.

19. Bureau of Reclamation, *Valves, Gates and Steel Conduits*, Design Standards No. 7 (1956).

20. Gomes Navarro, J.L. and Juan-Aracil, J.: *Saltos de Agua y Presas de Embalse*, Tipografia Artistica, (1964), Madrid.

21. Smith, L.G.: Floating-Ring Gate and Glory-Hole Spillway on Owyhee Dam, *The Reclamation Era* (Aug. 1940).

Chapter 2

Types of Gates

2.1 FLAP GATE

This type of gate consists of a straight or curved retaining surface, pivoted on a fixed axis at the sill. When designed with a leaf with the shape of a fish belly, it can be operated from one end for spans up to 20 m, since the closed shell structure offers high resistance against torsion. The bearings are rigidly anchored to the sill and spaced from 2.5 m to 4 m.

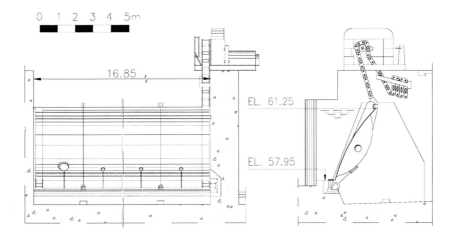

Fig. 2.1 Flap gate, Villeperrot Dam, span 16.85 m and height 3.3 m (ALSTOM)

In its fully raised position, the flap gate makes an angle from 60 to 70 degrees with the horizontal. In the fully lowered position, the skin plate forms a continuous surface with the weir bottom, presenting no obstacle to the water flow.

Similar to the drum and sector gates, the water flows over the flap gate when it is open. Seals are provided at the lower edge and the sides of the gate leaf. The lower seal may be made with a rubber strip bolted to the sill and the skin plate (Fig. 2.2), or with one end abutting on the cylindrical surface of the skin plate lower portion (Fig. 2.3). A cover plate usually protects the lower seal.

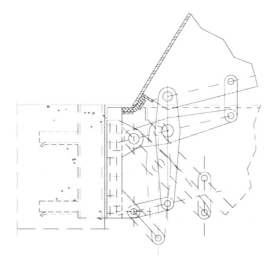

Fig. 2.2 Flap gate – Details of hinge and bottom seal (DSD-NOELL)

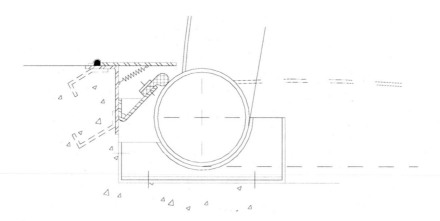

Fig. 2.3 Protection of the bottom seal

Floats may be incorporated into the hoisting mechanisms to provide automatic operation. Flap gates are often equipped with counterweights, to reduce the operating forces, as shown in Figure 2.4.

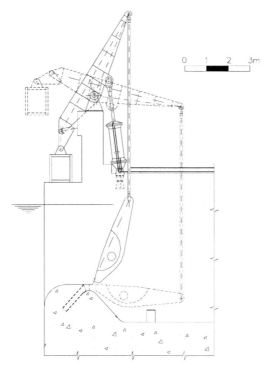

Fig. 2.4 Flap gate with counterweight, Juquiaguassu Dam (DSD-NOELL), span 7 m and height 3.5 m

Metal or rubber side seals, attached to the side end plates of the leaf, permit side sealing at all gate positions. If side sealing is desired only at the fully closed position, the seals may be mounted in the piers (Fig. 2.5).

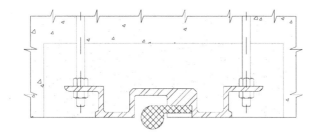

Fig. 2.5 Side seal mounted on end piers (RODNEY HUNT)

Flap gates operating partly opened may be subjected to severe oscillations, due to the creation of a low-pressure zone under the overflowing nappe. This problem can be eliminated or minimized through the construction of air vents with exits at the piers or with the aid of metal pieces installed on the gate leaf top, spaced from 2 m to 4 m, and designed for the purpose of breaking the stability of the overflowing water nappe (see Figures 2.6 to 2.8).

Fig. 2.6 Downstream view of a flap gate with nappe breakers (RODNEY HUNT)

Fig. 2.7 Flap gate, Ottendorf Dam (M.A.N.), 30 m wide by 5.5 m high

Flap gates are also used on the top of segment and fixed-wheel gates. These gates so formed present great operational flexibility, for they permit:

a) precise regulation of the reservoir level, through gradual lowering of the flap gate, which discharges with a low water head;

b) passage of ice and other floating material, by lowering of the flap gate, with little loss of the reservoir water;

c) discharge of a large water volume, by lifting of both gates.

Fig. 2.8 Flap gate, Thunn Dam (ZWAG), 12 m wide by 4.15 m high

Fig. 2.9 Segment gate with flap, 14.5 m wide by 7.7 m high, Wynau Dam (GIOVANOLA FRÈRES)

In this type of construction, the flap gate hinge bearings are fastened to the lower gate leaf and the skin plates of both gates are connected by a rubber strip screwed on to them, which assures the joint water tightness for any relative position of the gates. The flap gate is usually built with a smaller span than that of the lower gate: the later is provided with vertical plates (aprons) designed to make contact with the flap gate side seals at all gate positions.

Flap gates used as the main element of spillway control have their maximum height limited to about 5 m; yet they may be built for large spans, 50 m or more. The largest flap gate ever manufactured is that of St. Pantaleon, Austria, with a 100 m span and a 3.7 m height. The highest gate is that of Barenburg, Switzerland, with a 6.2 m span and a 7.2 m height.

2.2 CYLINDER GATE

The cylinder gate has a cylindrical-shaped leaf, which executes a vertical translation movement. Cylinder gates are generally intended for use in intake tower structures; they can be designed either for external or internal pressure.

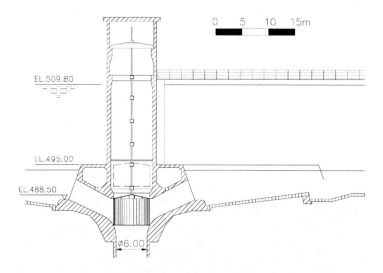

Fig. 2.10 Cylinder gate, Vianden Powerplant

Lifting of the gate uncovers lateral ports radially arranged in the tower, through which the water passes. Generally, the upper seal is screwed onto the embedded parts and contacts permanently the outer cylindrical surface of the gate, preventing water passage even for partial openings. The bottom seal is fastened on the leaf; when the gate is closed, the seal bears against the sill structure.

The cylinder gate is usually raised by a series of vertical stems, which extend upward to the tower top, where their ends are connected to motor-driven screw lifts. Due to the cylindrical shape of the skin plate and its vertical arrangement, the hydraulic pressure is balanced. So, for movement of the gate, it is sufficient to overcome the weight of the gate and stems, seals and guide friction and downpull forces arising during opening or closure.

Gate stems are guided by bearings fixed in the tower structure and spaced as required to prevent buckling when the gate is being forced down against friction.

An interesting application of the cylinder gate is that installed in 1936 by the Bureau of Reclamation in the morning-glory spillway at the Owyhee dam, Oregon, USA. The gate has an 18 m diameter and a 3.6 m height, and received the special name of ring gate. It is installed in a hydraulic chamber at the top of the spillway structure (see Figure 2.11). Upon lowering of the gate, water is admitted into the spillway, passing over the gate structure. The gate is raised or lowered by buoyancy in water introduced in the hydraulic chamber from the reservoir, through a 600 mm diameter inlet pipe. Water is drained from the hydraulic chamber through two 600 mm diameter needle-type control valves, which in turn are controlled by a system of a float and control cables and sheaves.

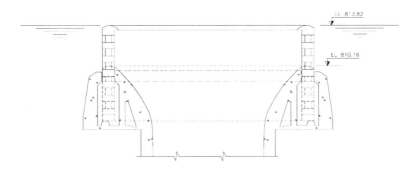

Fig. 2.11 Ring gate, Owyhee Dam, 18 m diameter and 3.6 m height

Metal seals are provided at the inner and outer lips of the hydraulic chamber. The seal on the inner side prevents the escape of water from the hydraulic chamber, and the seal on the outer side prevents entrance of reservoir water into the chamber. The gate is also provided with a vertical guiding device that keeps the gate level and, at the same time, prevents the gate from rotating in the chamber.

2.3 STOPLOGS

Stoplogs are primarily used for maintenance and repair of main equipment or gates. It can be used at:
. upstream of intake gates;
. upstream of spillway gates;
. upstream and downstream of bottom outlets, when the downstream water level is higher than the gate sill;
. downstream of turbines or draft tube emergency gates.

Their construction is similar to that of the slide gate: they do not have wheels or rollers, except in the infrequent and particular cases of sloped guides or of shutting-off the flow of water through lowering of the stoplog.

Depending on the height to be sealed, the stoplog may comprise more than one element. These are called stoplog panels. The height and quantity of stoplog panels are influenced by:

. the lifting capacity of the crane or gantry crane;

. gantry crane height - the higher the panel, the greater the lifting height above the operating deck;

. difficulty of storing high panels;

. transportation limits of the access ways to the site - the maximum normal width of parts to be conveyed by highway or railway is 3 m. In highways, if such limit is exceeded, a special transportation license is required, as well as for the use of forerunners, which increases transportation costs.

Only a careful evaluation of the influence of all these factors will allow the designer to determine the adequate panel height. It is worth noting that a large number of panels per opening demands also a longer time for complete installation or removal.

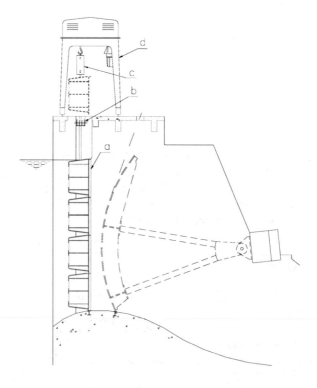

Fig. 2.12 Spillway stoplogs
(a) stoplog panel; (b) dogging device; (c) lifting beam; (d) gantry crane

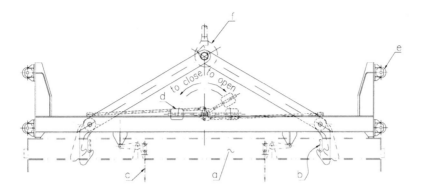

Fig. 2.13 Lifting beam
(a) stoplog panel; (b) lifting lug; (c) by-pass valve stem actuator; (d) counterweight; (e) roller; (f) hook

Weir stoplogs are provided with seals on their sides, sill and between the panels. The panels may be identical or designed such that each resists the corresponding pressure, in which case it is necessary to obey a defined sequence of installation. The option of identical panels is not, obviously, the most economical since all panels must be designed for maximum head. Their main advantages are greater safety, flexibility and quickness of operation, because there is no sequence to follow in the installation of the panels. In case of submerged stoplogs, the upper panel requires a seal on the top and so cannot be identical to the other panels.

Fig. 2.14 Lifting beam and stoplog panel, Porto Colombia Dam

Stoplog panels are placed and removed under balanced pressures, by means of cranes, traveling hoists or gantry cranes, with the help of lifting beams. The lifting beam is provided with lifting hooks operated by a system of levers and counterweight. It operates as follows:

a) to remove the panel - the lifting beam is suspended by the crane hoist and the counterweight is set to the position '*close*'. The beam is introduced into the guides and lowered till it hits the panel. The beam hooks then slide their sloped faces over the suspension lugs, rotating on their suspension pins until the hook ends penetrate the lug holes, when by action of the counterweight, the hooks regain the vertical initial position, fastening the panel.

b) to place the panel - with the panel suspended by the lifting beam, the counterweight is moved to the position '*open*'. The panel is introduced into the guides, being lowered until it reaches the lowest point of the guide length. Carrying on the lowering of the beam, the hook ends disengage from the lifting lugs and the counterweight commands the opening of the hooks, releasing the beam.

Fig. 2.15 Semi-elliptic spring for pre-compression of seals

After the gate placement, the water passage is drained. As the panels are placed under balanced head, it is necessary to mechanically press the seals against the seats in order to prevent passage of water to the region to be drained. Pre-compression of seals is obtained by semi-elliptic springs installed at the gate pressure side. The springs press against wedges fastened to the embedded parts.

The balance of pressures required for removal of the panels after conclusion of the maintenance services is achieved through filling the space between the stoplog and the main gate or equipment with water. For submerged gates (for example, at intakes or draft tubes), the following filling systems may be used:

. by-pass valve installed in the upper panel, operated by the lifting beam (see Figure 2.16); and

. flood valve embedded in the concrete structure.

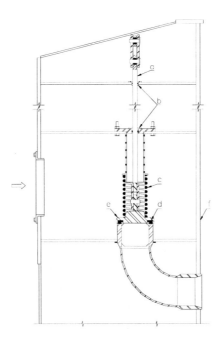

Fig. 2.16 By-pass valve
(a) stem; (b) bushings; (c) spring; (d) seal; (e) disc; (f) skin plate

In the particular case of weir stoplogs, the by-pass valve can be eliminated. Pressure balance is reached by cracking the upper panel, which creates a gap between this panel and the one immediately below. The comparatively low head acting on the upper panel facilitates this operation. This method is largely used in Brazil (for example, spillway stoplogs of the Estreito, Furnas, Porto Colombia and Itumbiara Power Plants).

Storing of stoplog panels may take place in the upper portions of the gate guides, near the operating deck (see Figure 2.17), or in recesses specially built in the concrete structure.

Fig. 2.17 Dogging device

2.4 SLIDE GATE

The slide gate is the simplest type of flat gate. It consists basically on a gate leaf that slides along side guides embedded or fastened to the concrete (Fig. 2.18). The leaf is provided with sliding surfaces, usually metallic, which under tight contact at the bearing surfaces act as seals.

Thanks to its simple and safe operation and because it requires little maintenance, the slide gate is largely used as a control device in irrigation canals, sewage works, bottom outlets and small intakes and reservoir spillways.

Other distinguished characteristics of the slide gates are the uniform transmission of the hydrostatic load to the concrete and the absence of vibrations in partial openings due to the large friction forces developed between the sliding surfaces. This feature is highly desired for bottom outlet gates. On the other hand, the slide gate is not recommended for installations requiring closure by gravity, due to the large friction forces created.

Slide gates may be made of timber, cast iron, cast steel or structural steel.

Fig. 2.18 Stainless steel slide gate (H. FONTAINE)

Stainless steel sluice gates with ultra high molecular weight polyethylene (UHMWP) seals are made by H. FONTAINE, Canada.

While metal gates are primarily recommended for their strength under high heads, the timber gates are economical, durable and have corrosion-resistant properties.

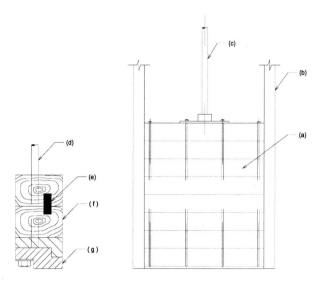

Fig. 2.19 Timber slide gate (RODNEY HUNT)
(a) gate; (b) side guide; (c) operating stem; (d) tie rod; (e) timber spline; (f) timber beam; (g) bottom seal

Timber gates up to 4 m by 4 m are used in weir works; sluice gates are usually designed for a maximum head of 7.5 m above the sill. The gate leaf consists of wooden horizontal beams, planed on all surfaces and rigidly keyed with metal or wood splines running the entire length of the timbers (Fig. 2.19). Vertical tie rods extend from top to bottom to hold the individual timbers securely together. Gate guides may be of cast iron, wood or structural steel and extend to a height equal to twice the gate height.

The leaf of cast iron gates is of one-piece construction. Seals are made of rubber, brass or bronze flat bars. Machined metal seals are fastened to the skin plate downstream face by countersunk head stainless steel screws. In the installation of screws, a minimum clearance of 1 mm between the sliding surface and the top of the screw head should be observed (Fig. 2.20).

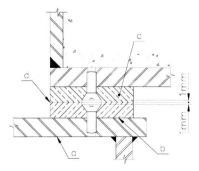

Fig. 2.20 Attaching of metallic seals
(a) skin plate; (b) seal seat; (c) metallic seal; (d) sliding surface

When seals and seats are metallic, materials of the same chemical composition should not be used in their manufacture. The seal material should be slightly softer than that of the seats, so as to avoid seizing of the mating surfaces under load and so that the wear take place on the seal and not on the seat.

Fig. 2.21 Cast iron slide gate (RODNEY HUNT)

Rodney Hunt Co., USA, manufactures cast iron slide gates, up to 4.8 m by 4.8 m, for seating heads of 24.2 m. and unseating heads of 15.2 m [1]. Seals and seats are formed of bronze strips locked in dovetail grooves machined in the cast iron frame (Fig. 2.22).

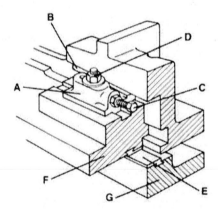

Fig. 2.22 Wedge system (RODNEY HUNT)
(a) bronze wedge; (b) bronze wedge seat; (c) bronze adjusting screw with lock nut; (d) hold down bolt; (e) bronze seat facing; (f) gate disc; (g) gate frame

On slide gates of welded construction, the leaf consists of a flat skin plate reinforced with upstream welded rolled sections; some manufacturers replace that construction by a corrugated skin plate with a trapezoid shape. Bronze or brass-machined bars attached to the downstream skin plate face usually provide sealing. Rubber or wood seals are also used. On gates with metal seals, the transmission of the hydrostatic load to the embedded parts is carried out through the seals.

To move the slide gate it is necessary to overcome, besides the gate weight and the downpull forces, the sliding friction between the seals (or the vertical end girders) and the embedded parts, either on raising or on lowering. Due to the high friction between those elements, the friction forces reach high values, almost always more than the gate weight. This means that the operating mechanism must provide a positive drive force on the descent, that is, must exert a downward force necessary to overcome the friction forces. So, the connecting stem located between the floor stand and the gate should be rigid; for the same reason, steel cable or chain hoists cannot be used. Lifting stems should be strong and guided by bearings spaced at regular intervals, to prevent buckling. Gates having a width equal to or greater than two times their height are provided with two lifting stems connected by a tandem shaft.

Slide gates are also recommended for use in high-head bottom outlets for control of discharges and prevention of silting. These gates must exhibit a reliable operation, free from vibrations, cavitation or instability. The combination slide gate-hydraulic hoist is very suitable to meet the desired features and assures safe operation of reservoirs. In such cases two identical gates are installed in tandem, spaced from 1 m to 1.5 m. The downstream gate stays normally closed and regulates the outflow; the upstream one, normally open, functions as a guard gate. While in this arrangement the upstream gate is not available for maintenance (except by completely emptying the reservoir), it presents, in relation to the design with an upstream stoplog, the possibility of shutting-off the outlet conduit at full discharge, which represents a guarantee of closure in the event of jamming of the regulating gate. Hydraulic hoists easily overcome the great friction force in slide gates of bottom outlets.

An interesting example of this type of installation is the bottom outlet of the Emosson dam, on the boundaries between France and Switzerland, near Chamonix. The main feature of these gates is the leaf construction. It was designed as a massive steel plate with no reinforcements. The advantages of this construction are the absence of any welding, the reduced size of the slots (which decreases the flow turbulence) and, finally, the large deadweight with its enhanced effect on dynamic load resistance.

The Emosson gates are 1.1 m wide by 1.8 m high and have a water head of 155 m. Side seals are made of bronze, with contact surfaces leveled with a file during assembling to assure good fit with the stainless steel seats. The gates are provided with a high-pressure lubrication system, which besides reducing sliding friction, allows detaching the gate from the guides after it has been pressed for a long period without moving. The rated pressure of the lubrication system is 50 MPa. The designers of these gates believe that, for economical reasons, the width should not exceed from 1.5 m to 2.5 m for pressure heads between 180 m to 60 m, respectively [2].

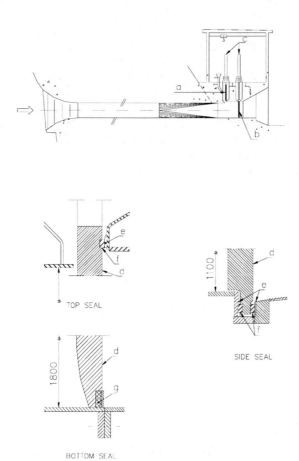

Fig. 2.23 Bottom outlet gates for the Emosson Dam (ZWAG)
(a) guard gate; (b) service gate; (c) hydraulic hoists; (d) gate; (e) stainless steel; (f) bronze; (g) rubber seal

2.5 CATERPILLAR GATE

A caterpillar gate consists primarily of a leaf supported by vertical girders at the sides. Continuous roller trains of the caterpillar type are mounted around the vertical girders. The roller trains travel with the gate.

The low friction on the rollers and their high load capacity recommend the use of the caterpillar gates in high-head installation requiring closure by their own weight, rather than fixed-wheel gates. The main disadvantages of this type of gate in relation to fixed-wheel gates are:
. higher initial costs;
. higher maintenance costs resulting from a great quantity of movable parts;

. the need of high precision adjustment between the roller tread and the roller tracks for
adequate operation;
. possibility of failure in some rollers or pins that may compromise the operation of the
roller train;
. possibility of damage to the rollers due to eccentric loading resulting from gate leaf
deflection, which requires the construction of horizontal beams of greater inertia.

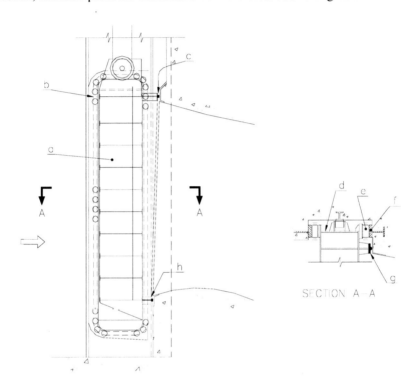

Fig. 2.24 Caterpillar gate
(a) gate leaf; (b) roller train; (c) top seal; (d) end girder; (e) roller; (f) roller track; (g) lateral seal;
(h) bottom seal

Caterpillar gates are generally designed as guard gates of high-head intakes and used
only in the fully open or closed positions except for filling the conduit in the cracking
operations and emergency closures. Normal opening and closure are carried out under
balanced pressures. Emergency closure for shutting-off the conduit is made by gravity.

The caterpillar gate comprises basically skin plate, horizontal beams, vertical end
girders, roller trains and seals. The skin plate may be placed upstream or downstream,
as well as the seals. An interesting design is the caterpillar gate of the San Louis dam,
USA, made in two sections, each with a pair of roller trains. The connection between
the two sections is made so that the horizontal splice exerts the function of a hinge to
permit slight deviations in the straightness of the tracks. This characteristic prevents the
possibility of high bending stresses in the vertical plane at the splice and helps to avoid
excessive roller loads, which could result from imperfections of the wheel track.

Fig. 2.25 Lower leaf of the San Louis Dam caterpillar gate (MITSUBISHI)

A special type of caterpillar gate was developed in the USA and received the name of Broome. This gate seats along its sides in a plane that is inclined relative to the plane of the tracks (Fig. 2.26). As the leaf motion is always vertical, any upward displacement separates the seals from the seats, thus eliminating the seal friction.

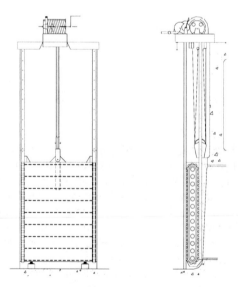

Fig. 2.26 Broome-type caterpillar gate

In Brazil, the following caterpillar gates have been installed:

Power Plant	Quantity	Span (m)	Height (m)	Head (m)	Manufacturer
Funil	3	4.5	6.25	77.8	ALSTOM
Marimbondo	8	6.6	11.44	37.3	Bardella
C.E.A.R.D.	2	2.8	4.40	60.0	Terni
Banabuiu	2	3.0	3.00	59.0	ALSTOM
Itaparica	6	9.5	11.00	38.0	BSI

Amongst the largest caterpillar gates already built, the following are noteworthy:

Year	Project	Span (m)	Height (m)	Head (m)	Outstanding feature
1938	Gènissiat	11.6	8.9	67.5	Span
1948	Tignes Les Brevieres	2.5	3.0	156	Head
1965	Guri	5.5	15.7	65/85.8	Height
-	Roseires	8.3	14.8	54.5	Area

2.6 MITER GATE

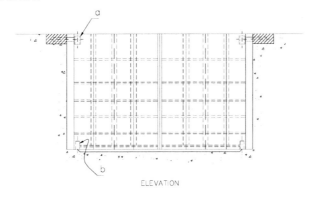

ELEVATION

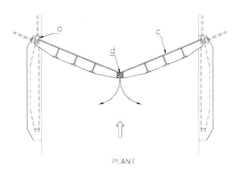

PLANT

Fig. 2.27 Miter gate
(a) gudgeon; (b) pintle; (c) skin plate; (d) center seal

The miter gate is used as navigation lock gates and comprises two rotating leaves with vertical hinge axes located in the lock chamber walls. In the closed position, the leaves meet at the center of the lock, supporting one another on the free ends, like a bishop's miter, hence its name. In the open position, the miter gate leaves fit into the recesses built in the chamber side walls (Fig. 2.28).

Fig. 2.28 Miter gate in open position, Bariri lock

The miter gate is fairly simple in construction and operation and can be opened and closed more rapidly than any other type of lock gate. Some disadvantages of the miter gate are:
a) submergence of the pintle bearing, bottom seal, and lower part of the gate, which require unwatering for inspection and maintenance;
b) susceptibility of the bottom seal to damage from debris on the sill;
c) extra length of lock chamber required to open the leaves (chiefly the downstream lock gate);
d) its inability to close off flow in an emergency situation.

Each leaf is usually supported at two points. A pin located near the leaf top and rigidly fastened to the gate passes through a collar bearing (gudgeon) anchored to the chamber wall. The lower part of the leaf presses against a thrust bearing (pintle), fixed to the sill. In certain cases, the lower support is made eccentric and designed so that the side seals move away from the seats when the movement of the gate opening starts.

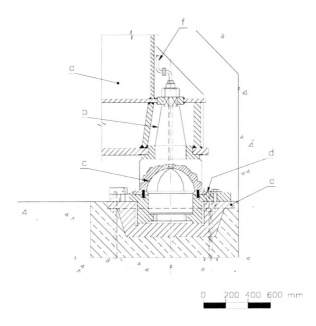

Fig. 2.29 Detail of the pintle of the Sobradinho lock miter gates (ALSTOM)
(a) gate leaf; (b) shaft; (c) semi spherical pin; (d) pin support; (e) pintle base; (f) grease pipe

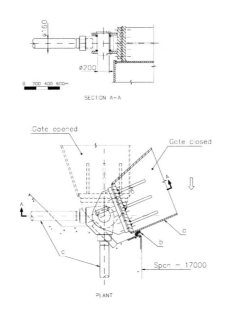

Fig. 2.30 Detail of gudgeon of the Sobradinho lock miter gate (ALSTOM)
(a) skin plate; (b) lateral seal; (c) anchors

Figures 2.31 and 2.32 show the forces acting on the gate and the bearing reactions. Figure 2.31 shows a plan of the miter gate bay. The reaction R_2 of leaf b on a is right-angled with the lock longitudinal axis. This fixes the point X of intersection of R_2 with the resultant P of the water thrust acting on leaf a. Reaction R_1 passes necessarily through the hinge axis and the point X. Knowing the direction of reactions R_1 and R_2, their magnitudes can be easily determined with aid of the next diagram.

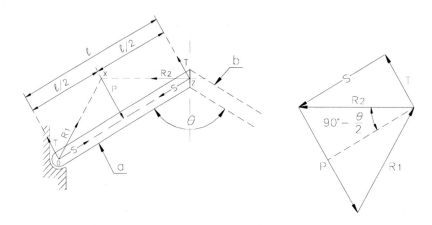

Fig. 2.31 Miter gate reactions, in plan

Figure 2.32 shows one of the leaves swinging free and supported on the bearings. Reaction P_1 is horizontal and defines point Y of the intersection of P_1 with the leaf gravity center axis. Reaction P_2 passes through point Y and the pintle center. The magnitudes of P_1 and P_2 are determined with the help of the diagram.

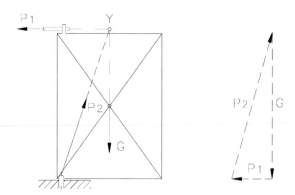

Fig. 2.32 Miter gate reactions, in elevation

The angle α between the leaf and the normal to the lock wall affects the structural stresses, the leaf length and, therefore, the gate weight. Usually, one takes (see Figure 2.33), A = L/4, which gives α = 26.565 degrees, since tan α = A/(L/2) = (L/4)/(L/2) = 0.5. Therefore,

α = arc tan 0.5 = 26.565 degrees.

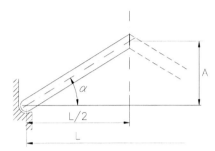

Fig. 2.33 Angle between gate leaf and lock wall

In most American lock gates, a variation of A in the range of L/6 to L/3 is noted, which corresponds to α from 18.4 to 33.7 degrees [3].

Miter gates are framed either horizontally or vertically. In the first case, which is the most usual arrangement, the entire water load is transmitted through the girders and bearings to the concrete walls. For vertical framing, part of the water load is carried by the top gate girder and transmitted to the wall and the other part is carried directly to the gate sill. Vertical framing is used only in wide gates with low height.

Leaves are provided with seals on the sides, center and the sill. Figure 2.34 shows the free end of one of the leaves of the Bariri lock gate, with a music note rubber seal and the center metal stops.

Fig. 2.34 Top view of central seal and leaf stops

Double-action horizontal hydraulic cylinders, attached to the leaf top girder operate most miter gates. The connection between the stem and the leaf is made near the hinge axis, in order to reduce the cylinder length. In normal conditions, the hydraulic cylinder exerts little force to move the gate for it is enough to overcome the friction forces.

2.7 ROLLER GATE

Roller gates are horizontal steel cylinders with toothed gears provided at each end. Racks are placed in the piers along the track recesses (see Figure 2.35). These gates were used with success in northern countries, where large masses of floating ice and low temperatures used to hinder or even block the operation of conventional gates. They are generally used on low head dams and in installations where a wide opening between piers is desired for the passage of ice or debris.

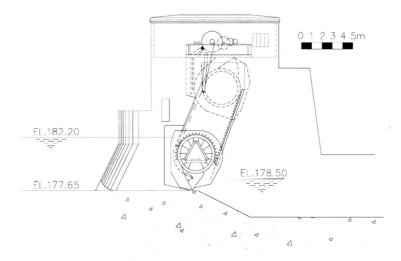

Fig. 2.35 Roller gate (KRUPP)

The cylinder is hollow and made up of curved steel plates stiffened by longitudinal members and diaphragm frames attached to the inside surface. The cylinder may be filled with water to prevent floatation. The gate is sealed at the ends and bottom. The bottom seal consists of a timber bolted to the bottom of the lip, on which the gate rests in the closed position. Timber or rubber strips attached to vertical shield plates exceeding the niche width provide end sealing. Side seats are made of plates or rolled sections embedded in the concrete. Metal discs connected to the cylinder ends make the transmission of water thrust to the embedded parts.

Because of the large resistance modulus inherent in the cylinder, the roller gate can be built for long spans. For the same reason, it can be operated from only one of its ends. Maximum practical dimensions of roller gates may reach 50 m in width and 8 m in height. It is probably the heaviest and the most expensive of all gate types.

The largest roller gates already built are:

Year	Project	Span (m)	Height (m)	Outstanding feature
1927	Ladenburg	45.0	6.5	Span
1938	Gallipolis	38.1	9.0	Height and area

Submersible roller gates (Figures 2.36 and 2.37) and roller gates with flap at their tops have also been built.

Fig. 2.36 Submersible roller gate, Kostheim Power Plant

Fig. 2.37 Detail of the roller gate gear rack, Kostheim Power Plant

2.8 SEGMENT GATE

The segment gate in its simplest form, consists of a curved skin plate formed to a cylinder segment, supported by radial compressed arms which transfer the hydraulic forces to fixed bearings (Fig. 2.38).

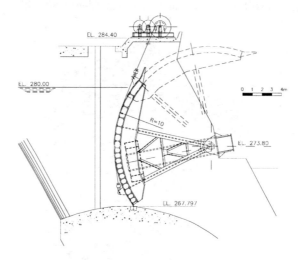

Fig. 2.38 Segment gate, Jupiá Power Plant, 15 m wide by 12.7 m high

The segment gate rotates about a horizontal axis, which passes through the bearing center and usually coincides with the center of the skin plate curvature radius. By this arrangement, the resultant thrust from the water pressure passes through the point of rotation and has no tendency to open or close the gate. In some cases, the center of curvature of the skin plate is located above the bearing axis so as to provide a lifting moment, which helps the winch in the opening of the gate. This moment must be less than that due to the gate weight to assure a positive closure. In certain installations, the gates may be counterweighted at the arms extension on the opposite side of the leaf to permit operation with very little power (Fig. 2.39).

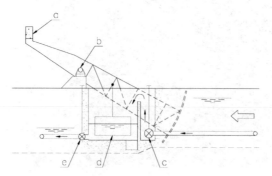

Fig. 2.39 Automatic segment gate with counterweight.
(a) counterweight; (b) bearing; (c) intake valve; (d) floater; (e) drain valve

It is very common for consultants, users and manufacturers of hydromechanical equipment to confuse the concepts of segment and sector gates. Often the segment gate is wrongly called a sector gate [4]. However, there is no reason for that mistake since the two types of gate have different features. Besides, half a century and an ocean separate their inventions: segment, 1853, France; sector, 1907, USA.

Thanks to its great simplicity and durability, the reverse segment gate (Fig. 2.40) is often used in lock aqueducts, as flow control valves in filling and emptying systems.

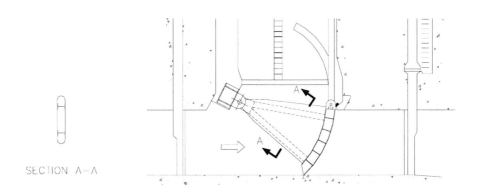

SECTION A–A

Fig. 2.40 Segment gate used in lock aqueducts

The first segment gates for lock hydraulic systems were oriented with arms subjected to compression. However, that construction caused the entrance of large volumes of air in the culvert through the gate shaft, causing excessive lock turbulence. Model tests indicated that reversing the gate position so that the skin plate concave side is faced to the upstream direction would eliminate this problem. Prototype experiences have substantiated this feature and reverse segment gates are now used almost exclusively in American locks [5]. These gates have small dimensions and their skin plate area does not exceed 25 m^2.

Large reverse segment gates have been installed, with success, in spillway works, in the region of Bavaria. Table 2.1 lists some of these gates, all supplied by M.A.N.. All these gates have a flat skin plate and are provided with flap gates on the leaf top.

Table 2.1 Reverse Segment Gates

Year	Project	Quantity	Span (m)	Height (m)	Area (m^2)
1965	Bittenbrunn	3	24.0	7.8	187.2
1964	Bertoldshei m	3	24.0	7.0	168.0
1966	Kleinosthei m	5	21.0	6.6	138.6
1956	Feldheim	3	16.0	8.2	131.2
1961	Offingen	3	19.0	6.6	125.4

The main advantage of the reverse segment gate is the introduction of loads into the piers by compression, which greatly facilitates their design.

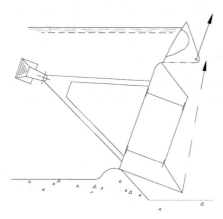

Fig. 2.41 Reverse segment gate with flap gate

The segment gate must be lifted to discharge. With a small opening, a comparatively large water flow is obtained, because the discharge is made under pressure, as an orifice. Gates intended to pass large quantity of floating ice or debris are usually equipped with a small flap gate at their top. With that construction it is possible to discharge debris or ice over the gate without appreciable loss of water. The flap gate is operated by a hydraulic hoist installed on the segment gate leaf (Fig. 2.42) or by chains or wire ropes.

Fig. 2.42 Segment gate with flap, Marchtrenk Power Plant (VA TECH Hydro)

Another type of construction that meets the need of both overflow and underflow characteristics is the double segment gate developed by ZWAG, Switzerland. It consists of two concentric leaves. The lower one has the conventional segment gate shape and may be lifted to allow large discharges. The upper one, placed upstream of the lower one, is provided with a pair of radial arms welded at the top and presses against the lower one by means of rollers.

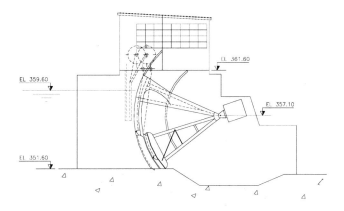

Fig. 2.43 Double-segment gate, Rupperswill-Auenstein Dam (ZWAG)

By lowering the upper gate, a precise regulation of the reservoir level can be achieved, as well as the passage of debris. The bearings of both upper and lower leaf arms are concentric and mounted on common shafts (Fig. 2.43). Both leaves can also be totally lifted, assuring free passage of the design flood.

As examples of the great versatility of the segment gates, one may quote the following applications:

a) submersible gates - used at upstream ends of small locks.

Their lowering permits the lock chamber filling and the passage of vessels. Two basic types are identified: one in which the hinge axis is below the sill (Fig. 2.44) and another, above (Fig. 2.45);

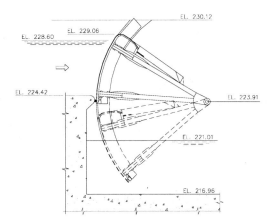

Fig. 2.44 Submersible segment gate, Saint Anthony Falls lower lock, Mississippi River

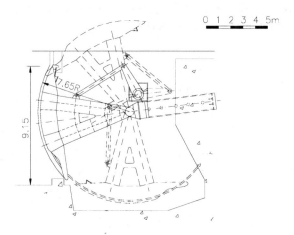

Fig. 2.45 Submersible segment gate, Nakaura lock (KURIMOTO)

b) gates with trunnions mounted on a vertical axis - used in pairs as lock chamber gates (Fig. 2.46). Gate leaves meet at the center of the lock when in the closed position. In the open position, the gates lodge in recesses built in the lock chamber walls. The chamber length of locks equipped with this type of gate is greater than that of locks provided with flat gates.

Fig. 2.46 Segment gate with vertical axis of rotation, Saint-Malo lock

c) automatic constant level gates - these gates are widely installed in irrigation canals, flood control works and on recreation lakes. They automatically maintain a constant water level without resorting to manual intervention or any other external power source. They include two types:
. constant upstream level gates;
. constant downstream level gates.

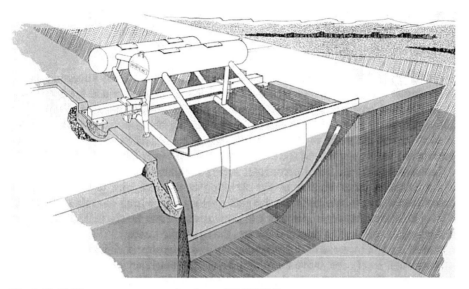

Fig. 2.47 AMIL constant upstream level gate (ALSTOM)

Gates for constant upstream level are always installed in weirs and comprise:
. a leaf with radial skin plate, which rotates around a horizontal axis;
. a float placed on the upstream face of the skin plate;
. a counterweight;
. arms and bearings.

In the design of such gates, two conditions have to be met:
- the gate trunnion axis, which coincides with the curvature center of the skin plate and the float, must be at the same elevation as the regulated upstream water level;
- the gate center of gravity must lie on a plane perpendicular to the lower face of the float and which also passes through the rotation axis.

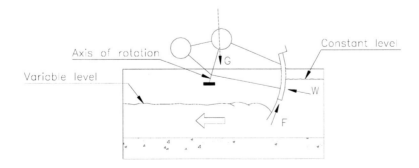

Fig. 2.48 Operating principles of the constant upstream level gate (ALSTOM)

The forces acting on the gate are:
. gate dead weight (including counterweights);
. water thrust acting on the submerged parts of the leaf and the float;
. bearing reactions.

For gate balance, it is necessary that the moment resulting from these forces in relation to the rotation axis be zero. The hydraulic forces on the skin plate and the cylindrical face of the float pass through the rotation axis and do not affect the equilibrium. In addition, the resistant moment produced by the bearing reactions can be disregarded, due to the small diameter of the gate pins. Thus, the balance depends only on the moments due to the gate weight and on the thrust on the lower face of the float.

The control of the upstream water level occurs as follows:
. if the upstream water level is at the elevation of the gate trunnion axis, the gate remains motionless, in any position; the gate opening depends on the rate of flow and the level difference between upstream and downstream water levels, that is, the greater the opening, the greater the flow and the less the hydraulic losses;
. in case of a slight increase of the incoming flow or downstream demand decrease, the upstream water level rises above the setting level. The buoyancy on the float increases and its moment exceeds that of the gate weight, causing the gate to open until the upstream level coincides with the elevation of the gate trunnion axis;
. in the reverse case, that is, decrease of the incoming flow on increase of the downstream demand, the upstream level is reduced; this causes a decrease of the upward thrust on the float and the gate closes until the upstream level coincides with the elevation of the trunnion axis.

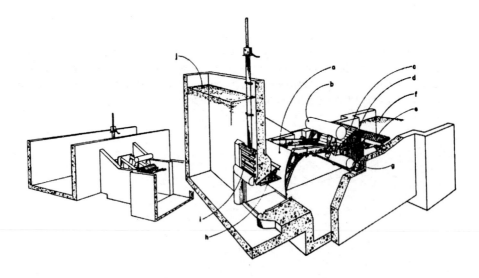

Fig. 2.49 AVIO constant downstream level orifice type gate (ALSTOM)
(a) gate leaf; (b) counterweight; (c) float; (d) float chamber; (e) metal walkway; (f) constant downstream level; (g) bearing; (h) metal lined opening; (i) emergency gate; (j) variable upstream level

With the gate open, the friction is limited to that of the bearings. On the other hand, in the closed position the skin plate edges may adhere to the canal walls and the opening of the gate requires an additional force, which is assured through a '*start float*' located on the leaf top, larger than the main float. The leaf is shaped to a trapezoid form to avoid gate jamming as the gate starts to open. As these gates are sensitive to the influence of wind and waves, they are usually provided with oil dampers to minimize the oscillations.

Gates for constant downstream level are designed for installation either on orifices discharging under pressure or on free-surface sluices. As in the constant upstream level gates, the gate hinge axis is located at the designed downstream level and coincides with the center of curvature of the skin plate and the float. Also, its center of gravity lies necessarily in the plane perpendicular to the lower face of the float and passes through the rotation axis.

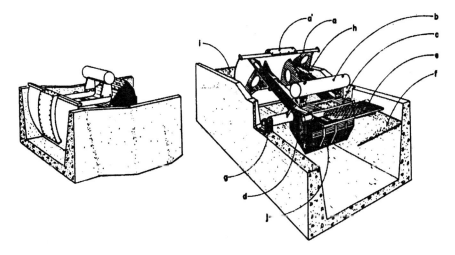

Fig. 2.50 AVIS constant downstream level surface type gate (ALSTOM)
(a) gate leaf; (a') damping tank; (b) counterweight; (c) float; (d) float chamber; (e) metal walkway; (f) controlled downstream level; (g) bearing; (h) embedded parts; (i) variable upstream level; (j) float-chamber communicating slot

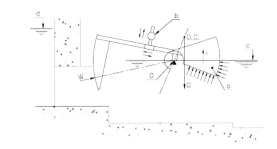

Fig. 2.51 Operating principles of the constant downstream level gate (ALSTOM)
(a) float; (b) adjustable counterweight; (c) constant downstream level ; (d) upstream level; (e) hinge axis

In these gates, the float is located downstream of the rotation axis, placed within a fixed chamber, which communicates with the downstream canal. With this arrangement, the float is not affected by oscillations generated in the canal. The operating principles are similar to that of the constant upstream level gates and are based on the balance between the moments caused by the gate weight and on the float buoyancy.

Its action on the flow makes it assume a certain definite position at which the flow required downstream can pass, the associated head loss being equal to the difference between the upstream level and the downstream level controlled at the region of the hinge axis. If the downstream demand decreases, the gate closes. If the demand increases, the gate opens, always keeping the downstream water level at the rotation axis elevation.

The sluice-type of constant downstream level gates is also fitted with a damping tank on the convex side of the leaf. Its bottom communicates with the upstream water through a wide opening whereas its top is connected to the atmosphere through a calibrated orifice. Constant level automatic gates have no seals.

The segment gate is the least expensive and the most adequate type of gate for passage of large floods, due to its simplicity of operation and maintenance (easy access to bearings and gate leaf framing), light weight and because it requires low capacity hoists. Hoists designed to overcome the dead weight of the movable parts and the friction forces on the bearings and side seals operate these gates. Gate closure is made by gravity. For greater safety of the equipment, the hoist is usually provided with a manual device to permit the gate lifting in case of power failure.

The absence of slots in the piers simplifies the design and favors the hydraulic flow conditions near the walls; this makes the segment gate suitable for operation in partial openings. When compared with flat gates, its installation is simpler. Erection and alignment of bearings and embedded parts do not require the time consuming inspection procedures of the embedded parts of flat gates, which require rigid erection tolerances of wheel tracks and guides, necessary for uniform load distribution.

Once the spillway maximum flow has been determined, the width of the bays should be selected so as to minimize the spillway length and therefore, its overall cost. It is usually desirable to use square gates or high gates rather than low ones. Gate bearings should be located at least 1 m above the water nappe flowing over the spillway crest under maximum flood discharge.

The bearing position affects the direction of the resultant of the water thrust acting on the gate. For reasons of spillway stability, it is desirable to have the direction of the resultant thrust approximately horizontal or directed downward, so as to reduce the overturning moment. Location of the bearings at a third of the gate height above the sill, causes the resultant to have a horizontal direction, while its positioning below that point directs it downwards. In practice, spillway segment gates have their bearings installed at a third to half the gate height. In the case of conduits and tunnels the trunnions are located above the water profile under free flow conditions.

The skin plate curvature radius affects inversely the gate weight, that is, the smaller the radius, the heavier the gate. On the other hand, the larger the radius and the higher the bearings, the higher will be the lifting required for passage of the maximum discharge. The radius is taken equal to the gate height or, at most, 20 per cent larger.

The choice of the sill position should be carefully made since it affects the gate height, the bearing location, the minimum lifting height and the crest pressures. By placing the sill upstream of the crest axis, the jet issuing under the gate will get detached from the concrete surface, resulting in low pressures and possible cavitation damage to the crest. With the sill placed downstream of the crest axis, the issuing jet is directed downwards and tends to follow the crest profile, usually resulting in positive pressures. The U. S. Corps of Engineers recommends that the sill should be placed between 1.5 m and 3 m downstream of the crest axis. In Brazil, this distance varies from 1.5 m to 4.5 m.

Table 2.2 lists a series of segment gates installed in Brazilian large dams with their main dimensions and the positions of the sill and the hinge axis.

Table 2.2 Geometric Data on Spillway Segment Gates

Project	Span B (m)	Height H (m) [1]	Radius R (m)	Hinge axis position E (m)	Sill position X (m)	Sill position Y (m)	H/B	R/H	E/(H+ y)
Itaipu	20.0	20.0	20.0	10.34	4.88	0.84	1.00	1.00	0.50
Tucuruí	20.0	20.0	20.0	9.07	4.30	0.47	1.00	1.00	0.44
Itumbiara	15.0	18.0	19.0	6.37	3.10	0.35	1.20	1.05	0.35
N.Avanhandava	15.0	15.0	15.0	6.69	1.70	0.19	1.00	1.00	0.44
Marimbondo	15.0	18.0	18.5	5.90	3.67	0.35	1.20	1.02	0.32
P. Colombia	15.0	14.7	15.0	5.00	2.13	0.20	0.98	1.02	0.34
Estreito	11.5	15.7	16.0	6.70	2.49	0.28	1.37	1.02	0.42
Capivara	15.0	15.0	15.0	6.50	2.20	0.16	1.00	1.00	0.43
Jaguara	13.5	18.0	21.0	8.50	4.50	0.50	1.33	1.17	0.46
Passo Real	11.5	11.0	12.0	4.77	1.99	0.27	0.95	1.09	0.42
Boa Esperança	13.0	12.5	14.5	6.46	1.82	0.21	0.96	1.16	0.51
Salto Mimoso	13.0	8.0	8.6	3.18	[2]	0.18	0.61	1.07	0.39
Jupiá	15.0	12.5	10.0	6.00	2.40	0.20	0.83	0.80	0.47
Limoeiro	12.5	9.0	10.0	6.54	1.48	0.15	0.72	1.11	0.71
Limoeiro II	11.0	6.5	8.0	2.66	1.45	0.16	0.59	1.23	0.40
Furnas	11.5	15.0 [3]	14.1	5.16	1.88	0.36	1.30	0.94	0.34

NOTES:
[1] Height above crest elevation, free board not included;
[2] Not available;
[3] In 1978 the gate height was increased to 17.5 m.

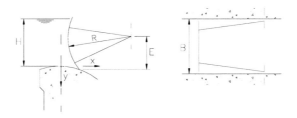

The skin plate supporting members are horizontal and vertical beams and stiffeners. In the construction based on horizontal beams and stiffeners, the beams bear on two main vertical girders, which in turn are supported by radial arms. In the construction with vertical beams and stiffeners, the main horizontal girders, also connected to radial arms, support these elements. The usual arrangements of the arms are shown in Figure 2.52.

The *a* arrangement is used in low gates or when the leaf frame is designed as a closed section resistant to torsion. The arms are also built with boxed sections with high torsional rigidity. In the *b* arrangement, the reactions on the supports of the vertical girders can be quickly determined, which is not the case of gates with three or more pairs of arms (cases *c* and *d*), in which the water thrust on the vertical girders and the support forces form a statically indeterminate system.

In addition, the design and the construction of the arms ends connection with the trunnion hub become rather complex in the case of gates with several arms, due to the reduced space, a reason why this construction is seldom adopted. Bracing of radial arms is made to prevent buckling in the vertical plane.

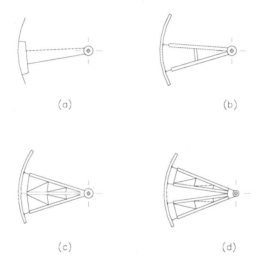

Fig. 2.52 Gate arms arrangement, in elevation

Fig. 2.53 Bottom outlet segment gate, Innerferrera Dam (ZWAG)

Fig. 2.54 Arms of segment gate, Furnas Dam

In the plane formed by the horizontal girders and the radial arms, the following designs may be used:

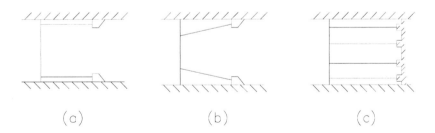

Fig. 2.55 Gate arms arrangement, in plan

In the *a* design, common in small gates, the arms are parallel to the pier face and connected to the ends of the horizontal beams. Wide gates are usually built with the *b* arrangement for it reduces the bending moment at midspan of the horizontal girders.

Sloping of arms creates an axial force in the horizontal girders, between the supports. The *c* construction, rarely used, is intended for gates subjected to large loads.

Fig. 2.56 Segment gate with inclined struts, Itaipu Dam

Fig. 2.57 Segment gate with four pair of arms, Haringvliet Dam, Netherlands, 56.5 m wide by 10.5 m high

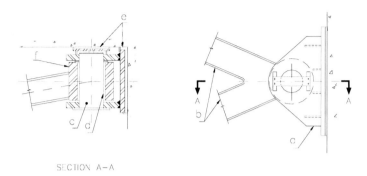

SECTION A—A

Fig. 2.58 Gate trunnion
(a) trunnion support; (b) gate arms; (c) pin; (d) bushing; (e) epoxy resin; (f) trunnion hub

The arms are usually fastened to the main girders with bolts and designed as columns subjected to axial loads and the bending moment due to bearing friction. In the axial direction, the arms are subjected to the water thrust and to the lifting force exerted by the hoists. The arms have a radial direction and are fastened to the trunnion hubs by welding or bolts. The hub is a hollow cylinder manufactured from cast or forged steel. A bronze bushing is fitted into the hub. The trunnion pins are made from stainless or forged steel; in this case, they are externally protected by a layer of hard chrome plating for greater corrosion and wear strengths. The pins are supported at both ends on the trunnion brackets and locked against rotation by a key plate. Trunnion yokes are usually cast steel or of structural steel and rigidly bolted to a concrete or a steel girder.

For bearings with bronze bushings, provision is made for a grease hole in the trunnion pins for lubricating purposes. Self-lubricating bushings are used in gates subjected to high loads, due to their low coefficient of friction and high compressive strength and, hence, greater load capacity. However, the grease lubrication is not always dispensed with in gates with self-lubricant bushings, for the grease occupies the gaps and hinders water and debris admission, preserving the original bushing lubricant.

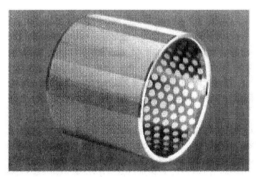

Fig. 2.59 Self-lubricating bushing (DEVAGLIDE)

Self-lubricating bushings are currently supplied under trade names such as Lubrite, Deva and Oiles. The lubricant is a mixture of several solid materials (metals, metallic oxides, and other materials) placed within radial holes machined in the bearing metal. Lubricants are joined to the bearing carrier material with heat and high pressure so as to fill the recesses with very densely packed plugs of lubricant. Bronze alloys, cast or stainless steels are used as base material for these bearings.

Gate bearings are usually subjected to very high loads and should be designed to compensate for the defects of pins misalignment, thermal elongation, elastic deformation and pillar inclination caused by settling. At the end of the 60's, the first spherical plain bearings were used in Europe, proving quite effective in the solution of these mentioned difficulties. The segment gates of the Nussdorf, Langenzrsdorf, Ferlach, Stör and Altenwörth dams are some examples of gates designed with spherical plain bearings [6]. These bearings comprise basically an inner and an outer ring with spherical sliding surfaces. The Stör segment gates have a width of 43 m and a height of 13 m and support a load of 37 MN. Here, spherical plain bearings with a bore diameter of 420 mm, an outside diameter of 770 mm and a width of 415 mm have been used.

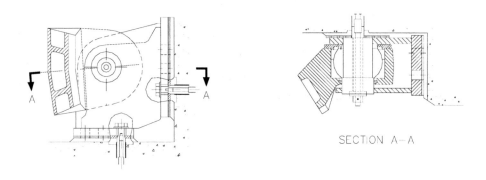

SECTION A–A

Fig. 2.60 Spherical plain bearing for segment gate, Stör Dam

On the convex surface of the inner rim a cage with round holes is welded into which are introduced sliding discs of glass fiber reinforced plastic containing PTFE. For assembly reasons, the outer ring is radially split. Sliding occurs between the projecting parts of the discs and the concave face of the outer ring, which is protected against corrosion by hard chromium plating. The free space in the housing at either side of the bearing is filled with a lithium base grease. These bearings were manufactured by SKF and designed for specific surface pressure of 90 MPa, under normal load conditions [7].

Segment gates are also fitted with roller bearings. This type of bearing has a low coefficient of friction, which enables the use of low capacity hoists. At this time, few applications are known; an outstanding example of its use occurred at the 34 segment gates of the Haringvliet dam, Netherlands, with a width of 56.5 m and a height of 10.5 m. The gates are mounted in pairs in the seventeen spans of the dam, arranged in opposite directions and supported on the main structure of the road bridge (Figure 2.57 and Figure 2.61). Every gate has four pairs of radial arms and, therefore, four sets of bearings.

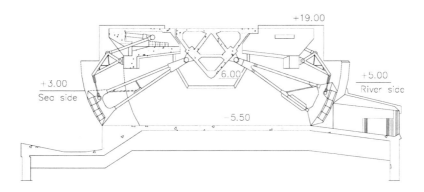

Fig. 2.61 Segment gates, Haringvliet Dam, Netherlands

The seaside gates are subjected to a water thrust of 16 MN, in normal conditions, and 60 MN, in case of a storm. Each pair of arms bears against two SKF 241/710 CA/C2 self-aligning spherical roller bearings (Fig. 2.62). The riverside gates are subjected to lower loads and have bearings SKF 241/500 CA/C2 [8]. In Russian dams the use of wood bushings in segment gate bearings is very common.

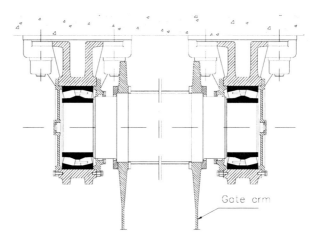

Fig. 2.62 Bearings of the Haringvliet Dam gates (SKF)

The thrust from hydraulic loading and part of the gate weight are transmitted from the trunnion yokes to the trunnion anchorage girders and thence into the piers. Two classic systems are used for anchorage of trunnions:
. structural anchorage; and
. prestressed anchorage.

The conventional structural anchorage comprises two longitudinal tension built-up members with a transverse anchor girder welded to their upstream members, as the means of transferring the hydrostatic pressures on the gate to the concrete by compression (Fig. 2.63).
The longitudinal members are isolated from the surrounding concrete by sheets of cork or similar material, placed during construction to permit their free movement or deformation, thus eliminating the tension stresses in the concrete.

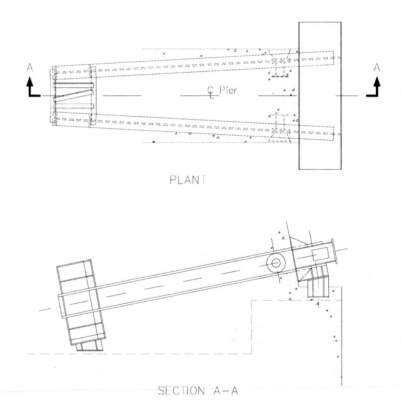

PLANT

SECTION A—A

Fig. 2.63 Built-up structural steel anchorage

The direction of the longitudinal members usually coincides with that of the maximum water thrust. In this system, the trunnions move slightly in the load direction due to the elastic deformation of tension members. Due to the evolution in size of segment gates, the structural anchorage became bulky, thus requiring larger piers. The prestressed anchorages were then conceived and developed for solution of this problem. This type of anchorage has greater loading capacity and makes possible savings in concrete since it permits the construction of narrower piers.

The prestressed anchorages have a number of post-tensioned round bars or tendons for fixing the trunnion yoke to the pier (Fig. 2.64). The tendons are longitudinally arranged in the pier in the same direction as the maximum thrust, inserted into pipes for isolation from the surrounding concrete. Tendons are fed into the pipes, tensioned by hydraulic jacks and anchored at the ends by wedges or cones. After prestressing is completed, the tendons are grouted.

Longitudinal prestressing should provide an average residual stress between the trunnion yoke and the pier, under maximum water thrust conditions. In some cases, the trunnion yoke is also formed from concrete and prestressed. References [9] and [10] may be useful for a better knowledge of the construction features of the prestressed anchorages.

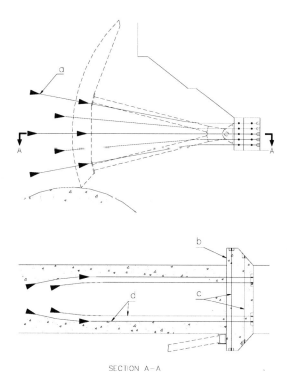

SECTION A–A

Fig. 2.64 Prestressed anchorages
(a) anchorage cone; (b) trunnion yoke; (c) transversal tendons; (d) longitudinal tendons

For gates with sloped arms, the side thrust should be taken into account in the design of the anchorages and in the determination of the bearing friction forces. To reduce friction due to the side thrust, it is usual to insert a thrust washer made of the same material as the bushing between the hub sliding face and the trunnion bracket.

The gap between the face of the pier and the trunnion bracket is usually filled with epoxy resin or molten lead, so as to transmit the side thrust to the pier.

Particular care must be taken in the design of the anchorages of the intermediate piers, since they must be dimensioned for two main loading cases:
a) one gate closed and the adjacent open;
b) both gates closed.

2.9 SECTOR GATE

The sector gate has a curved skin plate like that of the segment gate, but continued in its upper portion by a full surface, in the radial direction, up to the bearings, giving to the leaf profile the aspect of a circular sector (see Figure 2.65). This gate is hinged at the downstream side and its leaf has the shape of an open body on its lower radial side. The first designs of sector gates provided also the closure of the lower side, producing a closed section in shape of a real sector, hence its name.

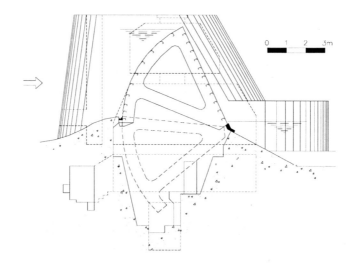

Fig. 2.65 Sector gate, St. Aldegund Dam (DSD-NOELL), 40 m wide by 5.4 m high

In the raised position, the sector gate is kept open by water pressure on the inner face of the upper radial side. Gate operation is fully hydraulic and hoists are not required. Its movable structure is placed in a gate chamber built in the crest structure. The reservoir water, when allowed to flow into the recess chamber, will cause pressure on the bottom of the gate, thereby rotating it upward. The gate is lowered through opening of the outlet valves, which drain the water held in the chamber. Hinge bearings are spaced from 1.5 m to 3 m, and fastened to the chamber downstream edge. The skin plate radius is taken as 1.4 to 2 times the gate height.

The sector gate provides a precise and safe automatic control without resort to an external power supply. Sector gates can be made as long as desired. Heights of gates are limited to about 8 m. This type of gate is regularly used in Europe, chiefly in sites where considerable quantities of floating material or ice need to be discharged over the structure.

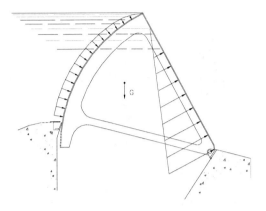

Fig. 2.66 Forces on a sector gate

Figure 2.67 shows the different manners of operation of segment and sector gates. Both are shown in a partially opened position. The segment gate must be lifted to permit discharge; with the gate slightly open the flow rate is comparatively high since the discharge is made in an orifice.

In the sector gate the discharge is made over the gate, by lowering it. Debris and ice are easily disposed of without appreciable loss of water.

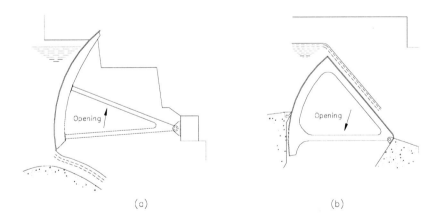

Fig. 2.67 Operating principles of segment (a) and sector (b) gates

In Brazil, the only installation of this type of gate occurred in 1924 at the Ilha dos Pombos dam, on the Paraiba river, where three sector gates were installed, with a width of 45 m, height of 7.4 m and radius of 10.6 m. The gate section has the shape of a hollow circular sector. Internal inspection of the gates is made through manholes located on the upper radial side.

Fig. 2.68 Downstream view of one of the sector gates at the Ilha dos Pombos Dam. Span 45 m

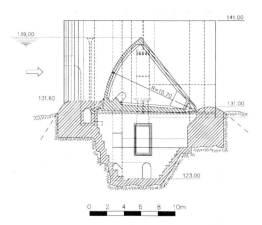

Fig. 2.69 Cross section of one of the sector gates at the Ilha dos Pombos Dam

2.10 *STONEY* GATE

This gate has a roller train on either sides of its frame, composed of horizontal rollers held in position by two vertical plates. The roller trains are placed between the gate leaf and the wheel tracks. The roller train is lifted by a steel cable running over a loose pulley installed on top of the vertical plates, with one end attached to an elevated fixed point on the pier and the other end being attached to the gate. With this arrangement, the roller trains move only one-half the distance covered by the gate during hoisting operations.

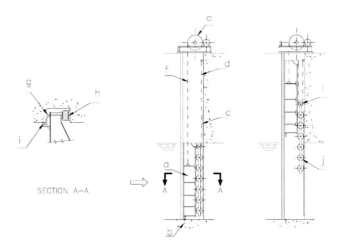

Fig. 2.70 *Stoney* gate
(a) gate leaf; (b) sill beam; (c) roller track; (d) roller train cable; (e) hoist; (f) gate leaf cable; (g) counterguide; (h) roller; (i) lateral seal; (j) roller train; (l) pulley

The main disadvantage of the *Stoney* gate is the presence of statically undefined reaction forces due to the application of numerous rollers and their variable positions in relation to the gate frame during traveling. In addition, with the gate raised the rollers get exposed to the full impact of the water flow, leading to appreciable wear of rollers by erosion due to the sediment content in water.

Fig. 2.71 *Stoney* gate with curved skin plate, Ilha dos Pombos Dam

Eight *Stoney* gates have been installed in the spillway of the Ilha dos Pombos Power Plant, Brazil, with a width of 12 m; two of them are 11.2 m high and six, 7.6 m high. Mechanical wire rope hoists, with concrete counterweights, operate these gates. The side seal consists of vertical steel pipes, which press simultaneously against the skin plate and the embedded parts. The skin plate is slightly curved and the leaf framing is made with trusses.

The largest Stoney gates ever manufactured are installed in spillways and have the following dimensions:

Year	Project	Span (m)	Height (m)	Outstanding feature
1925	Landenburg	40.0	8.3	Largest span
1929	Cize Bolozon	10.0	17.0	Greatest height

Because of its high cost and greater need of maintenance, this type of gate is no longer specified. The most recent application occurred in 1961, in the El Infiernillo dam, Mexico, where two *Stoney* gates were installed with a width of 3.3 m and a height of 8 m.

2.11 DRUM GATE

The drum gate leaf is a horizontal floating vessel, formed in the shape of triangular prism and hinged along its lower upstream edge (see Figure 2.72). Upstream and downstream sides are made of curved plates, whereas a flat plate forms the bottom of the gate. The structure is closed on the two sides by flat plates.

This type of gate is rather heavy but its operation is fully hydraulic and no hoists are required. All sides of the hydraulic chamber are sealed and the gate position is controlled by the application of headwater pressure underneath. Water supplied to the hydraulic chamber is controlled by slide or butterfly valves installed in special chambers built within the dam structure.

The fully lowered gate occupies the recess chamber of the spillway and the contour of its upstream face conforms to the ogee crest outline.

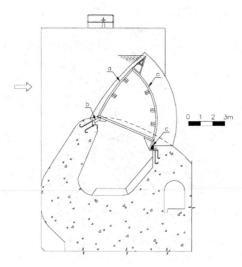

Fig. 2.72 Drum gate, Friant Dam, USA, 30 m wide by 5.5 m high
(a) skin plate; (b) hinge; (c) seal

The gate is kept in the fully closed position by the water pressure underneath the bottom side. The resultant water thrust on the bottom side creates a moment about the hinge axis capable of overcoming the moments produced by the water pressure acting on the upstream skin plate and its dead weight (see Figure 2.73).

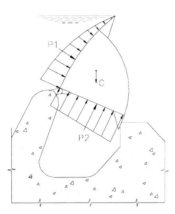

Fig. 2.73 Forces on a drum gate

The maximum gate opening is limited by a means of a metal stop. The water discharge is carried over the gate leaf.

Seals are installed in upstream and downstream lips and in the pier sides of the gate chamber so as to prevent leakage for any position of the gate. The upstream edge seal generally comprises a metal-spring plate fastened to the sill, with the free end contacting a curved plate in the extension of the upstream skin plate. A rubber or metal strip fixed to the sill seals the downstream edge. Its free end permanently contacts the downstream skin plate. Flexible drain hoses remove water that leaks into the interior of the gate.

By offering safety in case of floods, this type of gate is well adapted for crest spillways, since it does not require hoists or an external power supply for its operation. Also, the hinge reactions are uniformly distributed along the concrete edge and debris and ice can flow over the gate without considerable loss of water. The drum gates can be made as long as desired. The Hamilton dam gates, for example, have a width of 90 m and a height of 8.5 m. According to Hartung [11], however, its maximum height should be limited to 4 m; above that, the gate cost rises fast due to rapidly increasing use of steel for the gate itself and because of the large space needed for the hydrostatic chamber. Ordinarily, the arc of rotation of the gate is 70 degrees.

2.12 BEAR-TRAP GATE

The bear-trap gate is formed basically of two flat leaves, which are hinged horizontally at their lower ends. The free end of the upstream leaf presses continuously on the downstream leaf by means of rollers. The leaves form a broad inverted V in the raised and intermediate positions (see Figure 2.74).

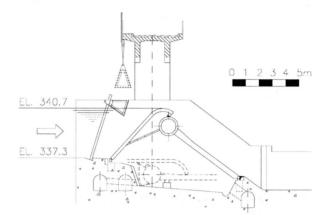

Fig. 2.74 Bear-trap gate, Schinznach-Bad Dam

The two leaves form a closed hydraulic chamber in conjunction with the base, connected with the upstream pool by conduits in the piers or the sill. The bear-trap gate is hydraulically operated and does not require hoists. Filling of the chamber causes a pressure to be exerted underneath the downstream leaf, which is provided with floating tanks. This leaf rises and pushes the upper leaf up with it. Emptying the chamber is achieved by connecting it with the lower pool and causes the lowering of the gate. The gate is sealed along its upstream and downstream hinged edges, at the juncture of the two leaves and against the piers at each end.

Bear-trap gates have been used in low-head installations as regulating gates in movable navigation dams and for log-sluicing operations. Its height is usually limited to 4 m. The principal advantage of the bear-trap gate is the considerable saving in the costs of substructure owing to the elimination of a deep hydraulic chamber.

Fig. 2.75 Bear-trap gate, Fröndenberg Power Plant, 15.5 m wide by 1.5 m high

The bear-trap gate requires a pressure of from 6 to 12 kPa under the lower leaf to begin its upward movement from the critical lowered position. The head available when the movable dam is down is insufficient. Therefore, other movable elements should be temporarily installed upstream of the gate to build up the head necessary to raise it. Compressed air can be used to aid in the initial raising operation.

The largest reported bear-trap gates are the seven 33.7 m wide and 5.5 m high gates installed at the Day Dam, in Vietnam. In the USA, up to 1948, bear-trap gates have been installed in pairs at all wicket dams on the Ohio River, except dams Nos. 21 and 48. These gates are up to 33.6 m wide and 4.6 m high.

2.13 FIXED-WHEEL GATE

The fixed-wheel gate is certainly the most common type of gate and consists basically of leaf, wheels, shafts and seals (Fig. 2.76). In general, the leaf is formed by a flat skin plate and stiffened by horizontal girders and ribs. On each side of the leaf, the ends of the horizontal girders are welded to a vertical girder. Wheels are mounted on shafts fixed laterally to the leaf on the vertical girders and have the double function of reducing the friction forces and the transmission of the water load to the embedded parts.

Fig. 2.76 Fixed-wheel gate for the bottom outlet of the Passo Real Power Plant

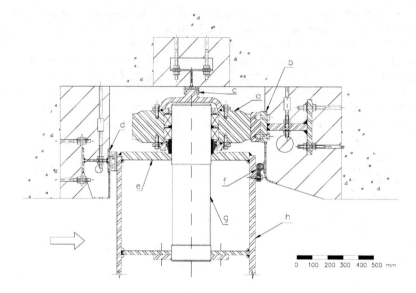

Fig. 2.77 Intake fixed-wheel gate, São Simão Power Plant (VOITH)
(a) wheel; (b) wheel track; (c) side guide; (d) counterguide; (e) end girder; (f) lateral seal; (g) wheel pin; (h) skin plate

The field of application of fixed-wheel gates is very large, as shown in Table 2.3.

Table 2.3 Typical Applications of Fixed-Wheel Gates

Project	Use	Span (m)	Height (m)	Head on sill (m)	Supplier
São Simão	Intake	6.50	11.3	43.78	Voith
Jurumirim	Bottom outlet	3.25	4.70	26.50	DSD-NOELL
Sobradinho	Lock gate	17.0	18.50	18.00	ALSTOM
Itumbiara	River diversion	5.00	7.86	26.15/88.15	Bardella
P. Colombia	Draft tube	10.85	5.43	35.6/47.0	ALSTOM
Bariri	Lock aqueduct	2.95	2.10	32.40	CKD Blansko
Bariri	Spillway	11.50	6.00	6.00	ALSTOM

Its main application, however, is made in installations requiring guard gates capable of closing under their own weight, as in high-head intakes. For the same reason, it is very common to use fixed-wheel gates as secondary devices for shutting off the flow of water in bottom outlets equipped with segment-type regulating gates.

Fixed-wheel gates used in weir installations are sealed at the sides and sill. The skin plate is placed either upstream or downstream of the frame girders. The upstream skin plate protects the beams against eventual damages caused by ice or trash accumulation, eliminates the downpull forces, reduces corrosion and facilitates maintenance. The skin plate extends from the sill to the reservoir maximum level, usually exceeding it by 0.3 m to 0.5 m, to prevent discharge or splash over the gate due to waves created by wind in the reservoir. Gate guides should extend above the operating deck in order to allow the gate to be lifted above the reservoir maximum level. A service bridge is installed at the top of the guides for housing the hoists. The sight of guides extending above the operation deck is generally considered a serious aesthetic drawback in this type of gate.

Fig. 2.78 Spillway fixed-wheel gates, Sameura Dam (HITACHI-ZOSEN), 10.4 m wide by 18.8 m high

Besides, the operation of fixed-wheel gates in spillways requires the lifting of all their dead weight, even for minor control of the reservoir level. Especially in the event of small lifts of the gate, foreign bodies can enter the gap provided for the flow causing damages to the bottom seal when the gate is being closed. These considerations constitute the reasons for using the following alternative designs:

a) multiple-leaf fixed-wheel gates - consists in the horizontal partition of the gate leaf into two or more sections, which can be independently driven. The passage of floating bodies and the accurate regulation of the water level are achieved by lifting the upper section;

b) double-leaf fixed-wheel hook gates - consists of two elements designed so that the upper one may be lowered, permitting discharge over the gate (Fig. 2.79). Both elements can be lifted above the maximum water level;

Fig. 2.79 Double-leaf fixed-wheel hook gate

c) fixed-wheel gates with flap - consists in the installation of a flap gate on the top of the fixed-wheel gate (Fig. 2.80). Lowering of the flap gate permits a precise regulation of the headwater level, as well as easy disposal of driftwood and ice.

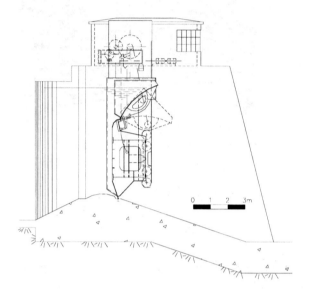

Fig. 2.80 Fixed-wheel gate with flap, Stockhein Dam (DSD-NOELL), 8 m wide by 6 m high

The double fixed-wheel gates were conceived in Europe, in the beginning of the last century. Ever since, three basic designs have been developed and are shown in Figure 2.81.

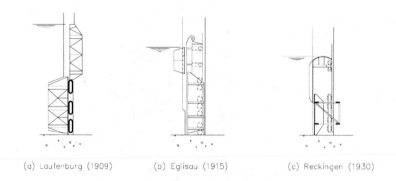

(a) Laufenburg (1909)   (b) Eglisau (1915)   (c) Reckingen (1930)

Fig. 2.81 Double-leaf vertical gates

In the Laufenburg gate the girders of the upper panel are placed downstream of the skin plate and the lower panel is installed in the reverse position. This design calls for separate wheel tracks, one for each panel, and may create hydrodynamic forces in the lower panel, which affects the hoist capacity. The Eglisau gate presents an inverted arrangement of the upper and lower panels in relation to that of the Laufenburg design. Although the hydrodynamic forces are eliminated here, the framework of the lower panel stands within the reach of debris and ice discharged over the gate. Furthermore, separate wheel tracks for each panel are still necessary, unless the vertical end girders of the lower panel are extended and used as tracks for the upper panel. The Reckingen gate was developed by M.A.N. and comprises a lower panel similar to that of the Eglisau gate and an upper panel with a single main girder at the top and wheels that move on the same wheel track used by the lower element. The bottom of the upper panel presses against the lower panel through small rollers. The leaf of the upper panel consists of a flat skin plate reinforced by upstream vertical stiffeners. The top portion of the skin plate is curved downstream and designed to protect the lower panel girders from the water discharged over the gate. As the outline of the upper panel resembles a hook, that gate received the name of hook gate.

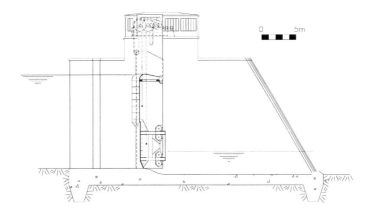

Fig. 2.82 Double-leaf fixed-wheel hook gate, Simbach Dam (THYSSEN-KLÖNNE), 23 m wide by 13.5 m high

Their main advantages may be summarized as follows:
. the hoisting forces required for the lower panel is reduced due to the absence of hydrodynamic forces;
. a single wheel track serves both panels.

High-head fixed-wheel gates are installed in recesses built in intake structures or in the face of concrete dams, at the upstream ends of outlets or penstocks. Since the primary use of these gates is to provide emergency shutoff of water to protect the penstock and downstream equipment, the preferable location of such gates is near the entrance. However, such gates are installed in recesses near the axis of earth-fill dams where the installation at the entrance is not feasible or too costly.

   The use of fixed-wheel gates with upstream skin plate and seals is common in Europe. Such preference derives from the absence of downpull forces in these gates, which involves a reduction of the hoisting forces and, consequently, lower costs. Upstream skin plate and seals also result in making it possible to inspect the gate framework when the gate is closed and the downstream side is unwatered. Special care should be taken in the design of the top girder of such gates so as to minimize its deflection under load, for leakage will occur if the upper seal is away from the upstream lintel.

Some examples of intake gates with upstream skin plates and seals found in Brazil are listed in Table 2.4.

Table 2.4 Examples of Fixed-Wheel Gates with Upstream Seals and Skin Plate

Project	Span (m)	Height (m)	Head on sill (m)	Supplier
Água Vermelha	9.53	10.40	40.0	Bardella
Capivara	8.50	12.13	58.0	Bardella
Foz do Areia	7.40	7.40	63.7	Bardella
Itaúba	3.80	6.26	22.5	Coemsa
N. Avanhandava	6.54	18.00	37.7	Bardella
Bariri	9.65	4.70	26.5	ALSTOM
Miranda	5.50	7.52	21.5	IMPSA

When an intake gate with an upstream skin plate and seals is used for filling the penstock, forces large enough to catapult the gate can develop. Several failures in installations of this type have occurred.

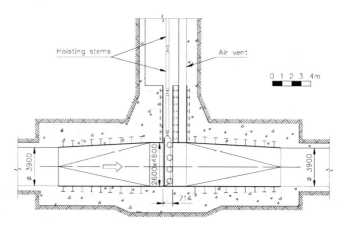

Fig. 2.83 Fixed-wheel gate, Uvac Power Plant, Yugoslavia. Head on sill of 59.5 m

At the Mossyrock Dam, USA, an intake gate weighing 654 kN was catapulted approximately 12 m up the gate slot. At the Dworshak Dam, a gate weighing 124.6 kN catapulted 76 m to the top of the gate slot [12]. At the UVAC Powerplant, Yugoslavia, an intake gate with a width of 2.6 m, a height of 4.6 m and a weight of 174.6 kN was catapulted when the tunnel filling was completed and the filling of the vent and the gate shaft started. The gate was damaged, some intermediate stems were twisted and the metallic air vent was flattened. At the time of the accident, the headwater level was 56.5 m [13].

At the Agua Vermelha Powerplant in Brazil, an intake gate was abruptly catapulted during the filling of the penstock, when a command for opening the gate was ordered, apparently without the pressure balance having been established. Filling of the conduit is done through a bypass valve installed in the concrete structure. The cylinder support beams were pulled out of the anchor bolts, the cylinder was partially thrown above the dam crest, water splashed through the shaft top opening and the gate leaf was damaged after falling on the sill [14]. The gate weighed about 1.4 MN and its main dimensions are shown in Table 2.4.

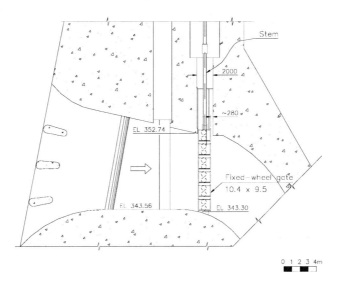

Fig. 2.84 Intake fixed-wheel gate, Água Vermelha Power Plant. Head on sill of 40 m

Model studies conducted by the U. S. Corps of Engineers led to the conclusion that the main cause of the two related accidents in the USA was that the width of the opening between the downstream side of the gate and the gate slot was smaller than the gate opening ($f < a$, in Figure 2.85). In such conditions, the back-of-the gate orifice could restrict the flow of water into the gate shaft enough to cause a hydraulic force directed upward to develop under the gate bottom. The Corps of Engineers suggested that an increase in the gap between the downstream side of the gate and the vertical shaft wall is beneficial in reducing uplift forces.

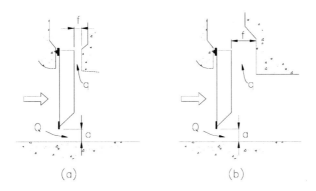

Fig. 2.85 Fixed-wheel gate with upstream skin plate and seals, in cracking position (penstock filling)

In general, model tests are not required for gates with upstream seals, as downpull forces are not developed in this kind of equipment. Nevertheless, the possibility of the development of uplift forces causing the gate to catapult reinforces the need to carry out model tests for upstream sealing gates.

Some transient hydraulic phenomena occurring with gates with upstream seals are very difficult to predict and can easily be observed in model tests, such as the contracted jet issuing vertically from the gap between the face of the gate and the upstream lintel. Although the gap is small, the local flow can be very high.

The sequence of events during the emergency closure of a gate with upstream seals is:
- Before starting closure, the gate shaft is full of water and the gate is completely submerged;
- As the gate starts to close, the pressure reduces downstream of the gate and the aeration pipe admits air;
- At a certain position of the gate travel, the water in the gate shaft is suddenly drawn to the low-pressure area located downstream of the gate. A large volume of water is retained at the top of horizontal girders. Draining of this water through the drain holes of the web girders is slow, as each girder drains over the next girder located below;
- Since the gate shaft is empty, the flow increases through the gap between the upstream lintel and the skin plate. The water jet flows vertically through the gap and falls over the gate top girder. The uncontrolled movement of the flow fills the interior of the gate leaf with water;
- Near the completion of the gate closure travel, the top seal contacts the upstream lintel and eliminates the gap and the local flow. The penstock becomes completely empty and the gate reaches the sill, ending the closure operation.

Intake structures with upstream sealing gate should not be designed with a heavy sloped portal invert, because of the high draining velocity of the water contained in the gate shaft during emergency closure. The higher the gate shaft drainage velocity, the higher the volume of water retained on the top of horizontal girders.

The velocity of re-feeding water entering the gate shaft can be controlled by appropriate design of the gap between the upstream lintel and the skin plate. The rate of flow will be reduced with a smaller gap. The weight of water retained in the interior of the gate leaf should be evaluated carefully and added to the gate weight, for estimation of the operating forces during emergency closure of upstream sealing gates.

The intake gate is the ultimate safety device for the turbine, and its main function is to close in an emergency, and in any condition of flow and water levels. From the point of view of safety, the most reliable equipment for emergency closure of intake structures is a gate with downstream seals, as it is not subjected to catapulting forces. Simulated studies show that the additional cost of the initial investment using gates with downstream seals instead of those with upstream seals as emergency devices for intake structures is very low, and should not be considered as the main factor in the selection of the seal arrangement [15].

The reliability of the equipment in closure operations should be the most important requirement in the selection of the type of emergency intake gate. However, gates with upstream seals have been preferred by designers only in relation to requiring a hydraulic cylinder of reduced capacity when compared with downstream sealing gates. This capacity is defined as a function of the forces acting on the gate during opening operations. In other words, the seal arrangement of intake gates is defined solely by the requirement of lifting capacity of the hydraulic cylinder for the opening operations, while the main function of this type of gate is to close in emergency conditions. It should be emphasized that the safety of the intake gate is more important than its cost, and preference should be given to emergency gates with downstream seals.

For gates having downstream skin plates and seals, downpull is always an important consideration, and particular attention should be given to the design of the gate bottom in order to reduce the hydraulic forces to a minimum. The downpull forces occur with the gate partly open, when a part of the head is converted into kinetic energy. The difference in pressures between the gate top and bottom develops hydraulic forces directed downwards, which act on the gate leaf. Such forces may exceed the gate weight and affect significantly the hoisting equipment. Downpull forces are examined in detail in Chapter 8.

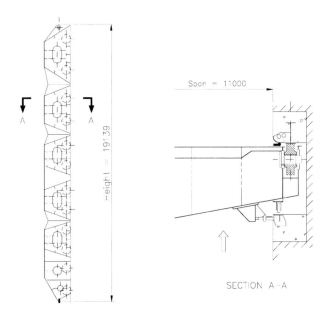

Fig. 2.86 Fixed-wheel gate, Sobradinho Power Plant (DEDINI)

The placement of more than two wheels on each side of the gate constitutes a statically indeterminate condition in the dimensioning of the vertical end girders. The study of the various hypotheses of loads on the end girders is rather complex, basically in what concerns the determination of the new reactions in the wheels in the same side of the gate, when one wheel goes off the wheel track. This may be caused by misalignment of one of the wheel treads, or by defects of erection or machining of the wheel tracks.

Contact of all wheels with the tracks can be ensured if the gate is designed as semi-flexible with a number of elements each fitted with only two wheels on either side (Fig. 2.87). Splices between elements are semi-rigid and designed so that they only transmit forces in the longitudinal direction of the skin plate.

Fig. 2.87 Roller gate leaf sections, Jupiá Power Plant (Bardella)

Splicing of elements is usually made with the aid of an overlaid plate, fastened by bolts. A rubber strip is placed between the skin plate and the connecting plate, for sealing purposes (Fig. 2.88).

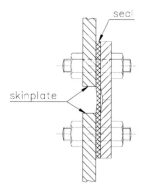

Fig. 2.88 Seal strip between gate sections

Submerged fixed-wheel gates have seals on all four sides. Seals on dam face gates frequently surround the opening like a picture frame. For gates installed in tunnels, the lower seal usually consists of a rectangular cross-section rubber strip, mounted either in the same plane as the side and upper seals or on the other side of the skin plate.

The largest fixed-wheel gates ever built were installed in the two navigation channels of the Volga Delta. Each gate weighs about 11.8 MN and closes a passage of 110 m span and 12.93 m high. Wire rope hoists with a capacity of 17.6 MN operate these gates [16].

2.14 VISOR GATE

Its name derives from the resemblance with the visors worn by the Middle Age knights in their helmets. The gate leaf has a box type structure with a semi-cylindrical skin plate, pivoted on horizontal pins (see Figure 2.89). The leaf is designed as a three-hinged arc.

Visor gates may be designed to seal in any direction. In the closed position, the leaf presses continuously against the sill. In the open position, it allows vessels to pass under the leaf, presenting no problem to navigation. The water thrust on the gate is transferred into the concrete structure through the pins fastened on the piers.

Opening of the gate is made by two mechanical hoists with wire ropes placed on concrete structures built on the piers. The closure is made by gravity.

Few applications are known of this type of gate. In the city of Osaka, Japan, three gates were installed in 1970, each with a clear width of 57 m and a height of 11.9 m, to control the tides. These gates are designed for a head of 10.9 m on the sea side and 6.7 m on the river side. The support length is 66 m and each gate weighs about 5.2 MN.

In 1960 six visor gates were installed near the cities of Hagenstein, Amerongen and Driel, on the rivers Lech and Low Rhine, Netherlands. The gates are 48 m wide and 7 m high. Gates are assembled in pairs and, similar to those in Osaka, are operated by wire rope hoists. To reach maximum opening (60 degrees), two hours and half are required [17].

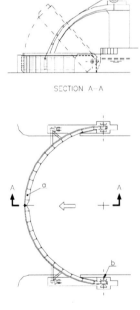

SECTION A–A

Fig. 2.89 Visor gate
 (a) gate leaf; (b) hinge

This type of gate is a much more economical solution for installations with very large spans than the usual construction with vertical lift or segment gates, and presents very favorable hydraulic features.

Fig. 2.90 Visor gate, Hagenstein Dam, Netherlands

REFERENCES

1. Rodney Hunt Co., *Flow Control Equipment Catalog*.
2. Schwarz, Hans-Joachim: Design Trends in the Construction of High Pressure Outlet Gates, *Swiss Dam Technique*, No. 21 (1970), Baden.
3. Cannell, P.J.: *Structural Design of Large Miter Gates*, Thesis presented to the University of Nebraska, USA.
4. Erbisti, P.C.F.: Segment or Sector Gate? (in Portuguese), *Construção Pesada* (Nov. 1979), São Paulo.
5. Douma, J., Davis, J. and Nelson, M.: United States Practice in Lock Design, *XXII PIANC* (1969), Paris.
6. Schuetz, K.: Vannes Segment Modernes et Leurs Paliers, *Schweizerische Bauzeitung*, 93 Jahrgang, Heft 42 (Oct. 1975).
7. Berthold, R.: Large Bearings used in Dam Gates, *La Revista de Rodamientos (SKF)*, No. 181 (1974).
8. Eylers, C.F. and Wiessing, J.M.: Paratoie Gigantesche Munite di Cuscinetti SKF, *La Rivista dei Cuscinetti a Sfere (SKF)*, No. 160 (1969).
9. O'Donnell, K.O.: Prestressed Concrete Anchorages for Large Tainter Gates, PCI Journal (June 1965).
10. Eberhardt, A. and Veltrop, J.: Prestressed Anchorages for Large Tainter Gates, *Proceedings of ASCE, Journal of the Structural Division*, Vol. 90, No. ST6, Paper No. 4170.
11. Hartung, F.: Gates in Spillways of Large Dams, *XI ICOLD Congress* (1973), Madrid.
12. Oswalt, N.R., Pickering, G.A. and Hart, E.D.: Problems and Solutions Associated with Spillways and Outlet Works, *XIII ICOLD Congress* (1979), New Delhi..
13. Djonin, K., Muskatirovic, J. and Predic, Z.: Some Experiences with Leaf Gate with Upstream Sealing, *XI Symposium of the Section on Hydraulic Machinery, Equipment and Cavitation*, IAHR, Proceedings, Vol.1, Paper No. 10 (1982), Amsterdam.
14. Ortiz, C.: Vertical Hydraulic Forces on Fixed-Wheel Gate with Upstream Seals during its Opening (in Portuguese), *VII SNPTEE*, No. BSB/GPH/05 (1984), Brasília.
15. Erbisti, P.C.F., Intake Gates: Upstream or Downstream Seals?, *The International Journal on Hydropower and Dams*, Issue 6 (2002).
16. Martenson, V.Ya. and Freishist, A.R.: Current State and Prospects of Development of Mechanical Equipment of Hydraulic Structures, *Hydrotechnical Construction*, No.12 (Dec. 1980).
17. Blokland, P., Die Kanalisierung des Nieder-Rheins, *Wasser U.Boden* (1960).

Chapter 3

Basis for Selection of Gate Type

3.1 INTRODUCTION

One of the most important tasks of a design engineer is the selection of a suitable type of gate for a hydraulic installation. As there is no established routine for this, selection should be based on a complete analysis of all factors capable of influencing performance, cost, quality and reliability of the equipment, such as:
- operational reliability;
- reduced weight;
- functional simplicity;
- ease of maintenance;
- advantageous structural requirements (slots, piers, gate chambers, guides etc.);
- magnitude and direction of forces transmitted to the concrete;
- gate hoist capacity;
- ease of transportation and erection.

Gate selection should always take into account past experience with successful types of gates as well as local manufacturing capability. In some cases it is evident that a great number of similar gate types exists in the same river or region. For example, reverse segment gates predominate on the Lech river and the German Danube river, double-leaf fixed-wheel hook type gates on the Austrian Danube river, fixed-wheel gates with flaps, double-leaf fixed-wheel gates and segment gates with flaps on the Inn river, and segment gates with flaps on the Enns and Drau rivers [1]. These groupings are almost certainly the result of local conditions or preferences.

3.2 MOST COMMON TYPES

Generally speaking, the most common types of gates used nowadays are:
- in intakes: fixed-wheel, slide, caterpillar, segment and cylinder;
- in spillways: segment, flap, fixed-wheel, sector, drum, segment with flap, fixed-wheel with flap and double-leaf fixed-wheel hook-type;

- in bottom outlets: slide, fixed-wheel, caterpillar and segment;
- as lock gates: miter, fixed-wheel and segment with vertical rotation axis;
- in lock aqueducts: slide, fixed-wheel and reversed-type segment.

3.3 OPERATIONAL REQUIREMENTS

In the selection of the gate type, it is important to adapt the characteristics of the gate to the operating requirements of the hydraulic structure. Some of the most significant requirements are described below.

a) Discharge capacity - in general, this is the most important feature for spillways. For large discharges, the option should be set on underflow type gates, such as segment and fixed-wheel gates. The capacity of the discharge is also affected by the geometry of piers and ogees. These structures should be free from obstacles such as lateral slots or projections. From this point of view, the best hydraulic efficiency is accomplished with gates that do not need lateral slots, such as segment, flap, sector and drum type gates.

b) Discharge of floating debris or ice - overflow type gates (flap, drum and sector gates) are widely used for precise water level regulation or for the purpose of passing debris or floating ice, with a small flow of water. Today there is a trend towards the use of double-leaf fixed-wheel gates or composite gates (segment or fixed-wheel gate with flap) to accomplish these requirements.

c) Headwater pressure operation - three types of gate are particularly adequate for operation without hoists: sector, drum and bear-trap. These gates are operated solely by the application of headwater underneath them. The headwater pressure creates a moment, which equals or overcomes the moments due to the gate weight and the frictional forces opposing the direction of gate travel. Water levels in the gate chamber can be regulated manually or by automatic float valves.

d) Loads on the concrete structures - water thrusts on gates are transmitted to the concrete through support hinges or tracks. These forces can reach very high values: each trunnion of the Itaipu spillway segment gates supports a load of about 21.6 MN. Design of the gate embedded parts is directly influenced by this item and may become determinant in the choice of a caterpillar gate in detriment to a fixed-wheel gate, for example. When compared with fixed-wheel gates, caterpillar gates present larger load capacity, due to their greater number of supports, which distribute the loading to the embedded track more uniformly.

Transferring the water thrusts on gates to the structures can be made:
- continuously, along the gate side guides - slide gates;
- through a series of supports, along the gate side guides: caterpillar and fixed-wheel gates;
- through a pair of trunnions located in the piers - segment gates;
- through a series of hinges attached to the sill - flap, sector, drum and bear-trap gates.

Special attention should be paid to the magnitude and direction of the forces transmitted by the gate, in the design of the gate supporting structures.

e) Absence of vibrations - it a very important characteristic for bottom outlet gates. The higher the friction forces between the gate and its guides, the less will be the tendency of vibration. Slide gates are widely used in bottom outlets due to the high friction resisting forces occurring between the slide gate frame and the seat facings. Segment and fixed-wheel gates are also recommended for outlet conduits.

f) Hydraulic regulation - segment, sector, drum, bear-trap and flap gates can be hydraulically regulated by means of floats and counterweights, with no hoists being needed. These types of gates are used to maintain constant pool levels near to non-flooding areas or for flood protection. They are also recommended for installation in localities of hostile climate (northern regions, for example) or of difficult access for operators.

Automatic control systems are somewhat complex and may be affected by the presence of sediment or debris in suspension in the water.

Automatic segment gates are widely used in irrigation, insuring upstream or downstream constant water levels.

g) Automatic closure in an emergency - intake gates are usually designed for automatic closure by gravity under unbalanced pressure conditions in case of power failure or turbine runaway or to shut down the water flow in case of penstock failure.

During the emergency closure the gate weight must overcome the friction resisting forces on seals and supports. Caterpillar and fixed-wheel gates are among the most used in intakes.

Gates designed for automatic closure by gravity should be necessarily operated by hydraulic hoists or mechanical hoists with cables or chains. Screw lift hoists cannot be used in this type of installation due to their self-locking characteristics.

3.4 PRESENT LIMITS OF GATE SIZES AND HEADS

Generally speaking, gate designers appear to be rather conservative in their approach, preferring whenever possible to adopt a type of gate already proven in practice, unless a new proposed type can be shown to have greater safety or economy. However, continuous development of hydroelectric projects and flood protection installations demands equipment of ever-increasing size, and designers are very often questioned about the maximum allowable dimensions and heads for a given gate type. Operating limits are constantly being extended. In a paper published in 1957 [2], Buzzel said: *"It is of interest to note that as late as 1946, it was firmly believed that no greater depth of water than 25 ft (7.62 m) could be discharged over an ogee spillway without shaking the structures to pieces..."*

Later, in 1962, G. Daumy [3] wrote: *"Par ailleurs, la vanne-segment èvacuateur de crue disposé en crete de barrage est dans la pratique, limitée à 16 m x 16 m."*
Nowadays these values are regularly exceeded. The spillway at Itaipu, for instance, has 14 segment gates of 20 m span and 21.34 m height, designed for a 20.84 m head over the sill (Fig.3.1). In its closed position, each gate is subjected to a water load of 43.3 MN (horizontal component). A 16 m x 16 m segment gate, as referred to by Daumy, would be loaded to only 20.5 MN. The Tucurui dam, also in Brazil, has 23 segment gates similar to those at Itaipu, 20 m wide and 21 m high. Several factors have influenced the increase of segment gate dimensions, such as the development of new welding techniques, use of high-strength low-alloy steels, prestressed anchorages (since 1956), self-lubricating bearings and hydraulic hoists [1].

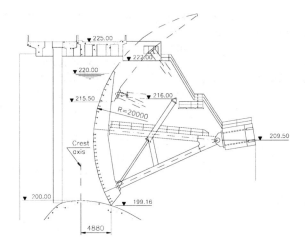

Fig. 3.1 Spillway segment gate, Itaipu Dam (DEDINI)

The author made a review of the state-of-the-art of hydraulic gates and the results are summarized in Tables 3.1 to 3.16. The gates listed in the tables are arranged in decreasing order of the hydrostatic load. For each type of gate, the tables present a series of the largest units selected among more than 4000 references supplied by manufacturers and contained in the bibliography consulted (see list at end of this chapter). The tables do not include gates for temporary use, e.g., for closure of diversion openings nor reverse segment gates, automatic segment gates with counterweights, double-leaf *Stoney*, neither submersible roller nor floating gates. For gates with curved skin plates, hydrostatic loads consist only of the horizontal component, because in most cases the skin plate radius and the position of the rotation axis are not known. The term *'high-pressure'* applies to gates with four-sided sealing, even in cases of gates submitted to relatively low heads.

Some of the listed gate types are no longer used, because of the high costs of manufacturing and maintenance, and the limitations of size and head. In these categories *Stoney*, roller and bear-trap gates are included. They are nevertheless presented here because of their wide use during an important development phase of hydraulic engineering.

For spillway gates the present trend is toward simplicity, such as the segment gate, the vertical lift gate, the flap gate, or the hydraulically operated drum gate. Selection may often be governed by factors such as the installed cost at the particular site, the type, quantity and size of debris, the power source available and in particular the reliability of operation. This last factor must consider simplicity of mechanism, site weather conditions (including freezing temperatures) and ease of access in an emergency [4].

Present limits of size and head for various types of gates are shown in Figures 3.2 to 3.5. The lines traced in the figures were based on the above-mentioned survey. The suggested limits cannot be considered conclusive but are representative of the present stage of development. These limits should be permanently updated. The main purpose of the figures is to outline the present fields of application of gates and to help the engineer in the preliminary selection of a gate type.

Figure 3.2 refers to high-pressure gates and shows gate area versus head. Span versus height of spillway gates are plotted in Figures. 3.3 to 3.5. It can be seen from Figure 3.2 that installations with heads above 175 m are equipped exclusively with slide gates but with restricted area (less than 20 m²). Fixed-wheel and segment gates are used approximately to the same extent, with a slight numerical superiority for the fixed-wheel type for the higher heads.

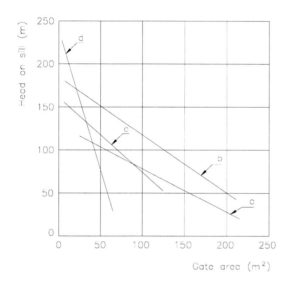

Fig. 3.2 High pressure gates - (a) segment; (b) fixed-wheel; (c) caterpillar; (d) slide

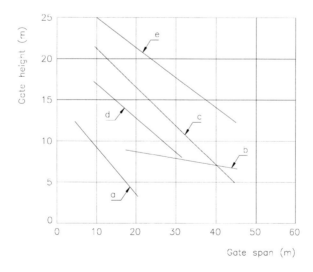

Fig. 3.3 Underflow spillway gates - (a) slide; (b) roller; (c) fixed-wheel; (d) *Stoney*; (e) segment

Figure 3.3 summarizes five underflow gate types. The preponderance of the segment gate over other types is clear. Also, the progressive increase in dimensions of slide, *Stoney*, fixed-wheel and segment gates, in that order, should be noted. This is undoubtedly because of a corresponding gradual reduction of the friction forces involved in these types of gates. *Stoney* and roller gates are no longer used and slide gates have their application restricted to small installations. Very few applications of spillway slide gates were found. Roller gates have been used in installations with very large spans, up to 45 m, but in no case was the gate height greater than 9 m.

Figure 3.4 shows four types of overflow gates. It is of interest to note that the curves of bear-trap and drum gates show an inversion of the general trend of decreasing height with increasing span, whereas sector gates have a neutral characteristic. This inversion may be justified by the fact that these types of gates are kept in the lifted position by the action of water pressure in the recess chamber. This pressure is obtained by a connection to the upper pool. As the pressure in the chamber increases with the gate height, this allows a larger gate span to be used. This situation does not occur with flap gates, which are moved by hoists and not by water pressure.

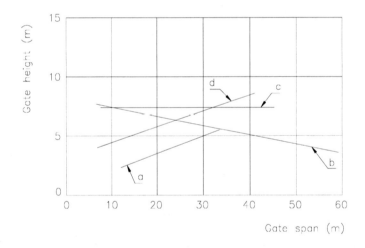

Fig. 3.4 Overflow spillway gates - (a) bear trap; (b) flap; (c) sector; (d) drum

Figure 3.5 shows three specially designed gate types. They can pass floating debris by lowering the upper leaf or the flap on the top of the gate, thus permitting small flows over the top of the gate. For larger discharges, both of these gate types may be completely raised. Double-leaf fixed-wheel hook type gates are very popular in Germany and France, and are designed with spans from 25 m to 40 m. Fixed-wheel and segment gates with flaps are generally designed for spans between 15 m to 45 m.

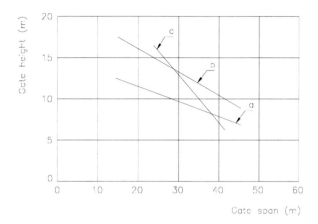

Fig. 3.5 Underflow and overflow spillway gates - (a) fixed-wheel with flap; (b) segment with flap; (c) double-leaf fixed-wheel

Table 3.1 Submerged Fixed-Wheel Gates

Year	Project	Quantity	Span (m)	Height (m)	Area (m²)	Head on sill (m)	Hydrostatic load (MN)	Manufacturer
1973	Tarbela	2	4.10	13.70	56.17	141.00	73.92	Sorefame
1983	Colbun-Machicura	2	5.80	13.40	77.72	102.40	72.96	VA TECH Hydro
1977	Guri	20	6.80	14.80	100.64	80.00	71.68	BYNSA
1979	Sobradinho	10	11.00	19.10	210.10	43.50	69.97	DEDINI
1963	Mangla	5	9.24	9.24	85.38	85.40	67.66	MAN
1979	Alicura	2	7.00	9.00	63.00	106.00	62.73	Riva-Calzoni
1994	Berke	2	8.50	15.50	131.75	55.00	61.07	VA TECH Hydro
1962	Malpaso	1	14.00	14.00	196.00	38.00	59.61	ALSTOM
1995	Ertan	4	4.00	8.00	32.00	190.00	58.39	VA TECH Hydro
2002	Tucurui	11	11.25	15.16	170.55	41.82	57.29	INEPAR
1983	Saddam	2	7.00	10.00	70.00	84.00	54.25	VA TECH Hydro
1979	Jebba	1	12.00	16.00	192.00	36.00	52.74	Mitsubishi
1963	Mangla	5	5.48	10.82	59.29	96.00	52.69	VA TECH Hydro
1975	Capivara	4	8.50	12.00	102.00	58.00	52.03	Bardella & others
1976	Tous	1	6.60	6.00	39.60	133.00	50.50	BYNSA
1994	Berke	1	7.00	10.00	70.00	71.00	45.32	VA TECH Hydro
1982	Tucurui	12	9.00	13.20	118.8	45.00	44.75	DEDINI
1979	Itumbiara	6	7.30	11.84	86.42	58.10	44.24	Krupp
1969	Ilha Solteira	10	8.50	9.00	76.50	58.50	40.53	ALSTOM
1968	Kastraki	1	6.00	8.50	51.00	85.00	40.40	VA TECH Hydro
1982	Itaipu	18	7.32	16.22	118.73	42.44	40.00	Bardella & others
1978	Água Vermelha	6	9.53	10.40	99.11	40.00	33.84	ALSTOM
1980	Foz do Areia	7	7.40	7.40	54.76	63.70	32.23	Bardella

Table 3.2 Submerged Segment Gates

Year	Project	Quantity	Span (m)	Height (m)	Area (m²)	Head on sill (m)	Hydrostatic load (MN)	Manufacturer
1965	Mangla	9	10.97	13.00	142.61	48.50	58.76	Krupp
1994	Berke	2	10.00	10.30	103.00	63.00	58.45	VA TECH Hydro
1972	Tarbela	4	4.88	7.30	35.62	135.60	46.11	VA TECH Hydro
1965	Tweerivieren	2	8.38	5.18	43.41	103.48	42.96	Kure
1981	Tabka		5.50	12.00	66.00	67.00	39.50	
1969	Cabora Bassa	8	6.00	7.80	46.80	82.30	35.99	Sorefame
1970	Reza Shah Kabir	4	8.00	6.70	53.60	71.50	35.83	ALSTOM
1979	Jebba	6	12.00	9.50	114.00	36.00	34.95	Mitsubishi
	Wuqiuangxi	1	9.00	12.00	108.00	38.7	34.64	
1962	Roseires	5	6.00	11.30	67.80	55.30	33.02	ALSTOM
	Toktogul		5.00	6.00	30.00	112.20	32.14	
	Nourek		5.00	6.00	30.00	110.00	31.49	
	Sayano-Sushenskoe		5.00	5.50	27.50	116.70	30.74	
1949	Castelo do Bode	2	14.00	8.50	119.00	30.00	30.06	ALSTOM
1981	Magat	2	6.00	12.50	75.00	46.50	29.61	VA TECH Hydro
1991	Aguamilpa	6	12.00	19.34	232.08	22.40	28.98	VA TECH Hydro
1962	Roseires	7	10.00	13.20	132.00	27.50	27.06	ALSTOM
1972	P. K. Le Roux	4	15.00	9.00	135.00	23.00	24.50	Sorefame
1974	Saddam	2	5.00	7.50	37.50	68.00	23.64	VA TECH Hydro
1960	Garrison	3	5.49	7.49	41.12	58.20	21.97	
1948	Chastang	2	13.60	9.50	129.20	22.00	21.86	ALSTOM
1974	Sobradinho	12	9.80	7.50	73.50	33.87	21.72	VOITH
1960	Mechra-Klila	4	16.00	12.30	196.80	16.90	20.75	ALSTOM
	Miranda	4	24.00	8.73	209.52	14.00	19.80	ALSTOM
1965	Cachi	1	6.50	4.30	27.95	74.00	19.70	Waagner-Biro
1961	Khashm El Girba	7	7.00	7.30	51.10	42.5	19.48	Riva-Calzoni

Table 3.3 Submerged Slide Gates

Year	Project	Quantity	Span (m)	Height (m)	Area (m²)	Head on sill (m)	Hydrostatic load (MN)	Manufacturer
	Sayano-Shushenskoe		5.00	8.45	42.25	135.00	54.20	
	Toktogul		5.00	7.00	35.00	112.20	37.32	
1979	Inguri	1	Dia. 5.00		19.63	181.00	34.86	
1996	Okutataragi	2	4.80	4.80	23.04	146.30	32.52	VA TECH Hydro
	Le Sautet	1	4.00	7.00	28.00	117.00	31.18	ALSTOM
1968	New Don Pedro	1	4.11	6.88	28.28	93.50	24.98	VA TECH Hydro
1995	Houay Ho	3	6.00	6.00	36.00	67.00	22.60	VA TECH Hydro
1979	Shintakase	2	6.40	6.30	40.32	60.00	22.49	Mitsubishi
1965	Nanairo	1	7.10	11.60	82.36	31.80	21.01	Kure
1964	Beaver	1	2.00	3.18	6.36	285.90	17.74	
1974	Niikap	1	5.40	5.40	29.16	63.70	17.45	Mitsubishi
1969	Azumi	2	4.90	4.85	23.77	76.00	17.15	Mitsubishi
1993	Huites	2	3.00	4.50	13.50	130.00	16.92	VA TECH Hydro
1980	Ulla-Forre	1	6.00	5.00	30.00	59.00	16.63	Kvaerner
1956	Flumendosa Mulargia	1	2.00	4.00	8.00	210.00	16.32	Riva-Calzoni
1956	Schwarzach	1	2.90	2.90	8.41	180.00	14.73	VA TECH Hydro
1970	Mica	9	2.29	3.50	8.02	189.00	14.72	VA TECH Hydro
1952	Saryar	6	5.20	5.80	30.16	45.00	12.46	Krupp
1960	Occhito	2	4.30	5.00	21.50	56.20	11.33	ATB
1974	Malta	1	4.20	4.60	19.32	55.00	10.00	ZWAG
1960	Arimine	2	4.70	4.40	20.68	47.20	9.13	IHI

Table 3.4 Submerged Caterpillar Gates

Year	Project	Quantity	Span (m)	Height (m)	Area (m²)	Head on sill (m)	Hydrostatic load (MN)	Manufacturer
1983	Colbun-Machicura	2	5.80	16.10	93.38	94.50	79.19	VA TECH Hydro
1996	Pangue	2	5.80	13.00	75.40	107.00	74.34	VA TECH Hydro
1965	Guri	14	5.50	15.75	86.63	85.80	66.22	Krupp
1954	Picote	1	8.00	10.00	80.00	87.00	64.35	ALSTOM
1938	Genissiat	2	11.60	8.90	103.24	67.50	63.86	ALSTOM
	Roseires	1	8.30	14.78	122.67	54.50	56.69	Sorefame
1988	Pehuenche	2	4.90	11.80	57.82	100.00	53.37	VA TECH Hydro
1972	Guri	20	6.00	9.70	58.20	90.00	48.62	BYNSA
1969	Cabora Bassa	5	8.90	11.00	97.90	52.00	44.66	Sorefame
1978	Tedori I	1	6.50	10.00	65.00	75.00	44.64	IHI
1978	Ayvacik	2	8.40	8.40	70.56	68.50	44.51	IHI
1966	Alcantara	4	5.50	8.60	47.30	94.20	41.71	ALSTOM
1966	Villarino	1	5.50	7.50	41.25	105.00	40.97	ALSTOM
1980	Odo	2	8.50	8.60	73.10	60.00	39.94	Hitachi-Zosen
1957	Derbendi-Khan	2	5.45	9.49	51.72	80.00	38.18	Krupp
	Nurek		3.50	9.00	31.50	120.00	35.69	
1969	El Chocón	6	10.00	10.00	100.00	40.40	34.73	Sorefame
1960	Miboro	1	5.50	8.60	47.30	77.00	33.73	IHI
1965	San Luis	4	5.34	6.99	37.33	92.72	32.67	Mitsubishi
1975	Aguieira	1	6.60	6.70	44.22	78.00	32.38	Sorefame
1965	Vouglans	2	6.00	9.00	54.00	64.00	31.52	ALSTOM
1972	Valeira	6	8.75	14.00	122.50	32.00	30.04	Sorefame
1974	Aguieira	3	5.00	7.50	37.50	83.30	29.26	Sorefame

Table 3.5 Spillway Segment Gates

Year	Project	Quantity	Span (m)	Height (m)	Area (m²)	Hydrostatic load (MN)	Manufacturer
1994	Wuqiunangxi	9	19.00	23.00	437.00	49.30	
1963	New Bedford	2	28.40	18.00	511.20	45.13	
1982	Itaipu	14	20.00	21.34	426.80	44.67	DEDINI & others
1982	Tucuruí	23	20.00	21.00	420.00	43.26	DEDINI
1950	Maxweel	5	25.60	18.30	468.48	42.05	
	Vilyui		40.00	14.00	560.00	38.46	
1971	Cedillo	3	18.30	20.30	371.49	36.99	ALSTOM
1976	Stor	2	43.00	13.00	559.00	35.64	Krupp
1964	Bekhme	3	19.00	19.45	369.55	35.26	
1970	Reza S. Kabir	3	15.00	21.28	319.20	33.32	ALSTOM
1979	Salto Santiago	8	15.30	21.00	321.30	33.10	Ishibras
1965	Guri	9	15.24	20.83	317.45	32.43	Krupp
1974	Salto Osório	9	15.30	20.77	317.78	32.37	Kurimoto/Confab
1973	Tsengwen	3	15.00	20.82	312.30	31.89	Kurimoto
1972	Valeira	5	26.00	15.73	408.98	31.56	Sorefame
1977	Itaúba	3	15.00	20.3	304.50	30.32	Bardella
	Fratel	6	18.75	18.00	337.50	29.80	Sorefame
	Carrapatelo	6	26.00	15.23	395.90	29.58	Sorefame
1964	Wanapum	12	15.20	19.80	300.96	29.23	
1967	Portage Mountain	3	15.25	19.50	297.38	28.44	VA TECH Hydro
1974	Agua Vermelha	8	15.00	19.60	294.00	28.26	ALSTOM
1966	Paldang	15	20.00	16.75	335.00	27.52	ALSTOM

Table 3.6 Spillway Slide Gates

Year	Project	Quantity	Span (m)	Height (m)	Area (m²)	Hydrostatic load (MN)	Manufacturer
1950	Salto Grande	8	12.76	9.50	121.22	5.65	ALSTOM
	Monsin	6	7.85	10.30	80.86	4.08	ALSTOM
1996	Pangue	1	5.80	11.50	66.70	3.76	VA TECH Hydro
1955	Kojima	12	12.00	7.50	90.00	3.31	Mitsubishi
1956	Serikawa	3	10.00	7.20	72.00	2.54	Mitsubishi
1955	Kuwanouchi	4	8.50	7.30	62.05	2.22	Hitachi-Zosen
1951	Hiraoka	2	7.30	7.53	54.97	2.03	Hitachi-Zosen
1974	Mihame	2	3.50	10.50	36.75	1.89	Hitachi-Zosen
1971	Takahama	4	5.40	8.30	44.82	1.82	Hitachi-Zosen
1963	Thanneermukkom	93	12.00	5.50	66.00	1.78	Jessop
	Sanga	1	12.00	5.30	63.60	1.65	ALSTOM
1976	Nabara	1	6.50	7.20	46.80	1.65	Hitachi-Zosen
1958	Peggau	1	12.00	4.80	57.60	1.36	Waagner-Biro
1968	Hallein	1	10.00	5.20	52.00	1.33	Waagner-Biro
1996	Besai	1	7.00	6.10	42.70	1.28	VA TECH Hydro
1919	Alzwerke Hirten	2	20.00	3.60	72.00	1.27	MAN
1957	Izawa II	9	6.00	5.90	35.40	1.02	Hitachi-Zosen
1929	Shannon	1	18.00	2.70	48.60	0.64	Thyssen0Klönne
1919	Alzwerke Hirten	2	9.70	3.60	34.92	0.62	MAN
	Rosières	1	10.80	1.70	18.36	0.15	ALSTOM

Table 3.7 Spillway Fixed-Wheel Gates

Year	Project	Quantity	Span (m)	Height (m)	Area (m²)	Hydrostatic load (MN)	Manufacturer
1979	La Grande	16	12.20	20.20	246.44	24.42	
1976	Funagira	9	20.00	15.30	306.00	22.96	IHI
1964	Ikehara	4	15.00	17.15	257.25	21.64	Mitsubishi
1973	Sameura	6	10.40	18.80	195.52	18.03	Hitachi-Zosen
1945	Keswick		15.24	15.24	232.26	17.36	
1956	Davis	3	15.24	15.24	232.26	17.36	
1960	Mihoro	1	14.00	15.80	221.20	17.14	IHI
1946	Belver	12	17.00	14.15	240.55	16.70	ALSTOM
1974	Yodogawa	2	55.00	7.80	429.00	16.41	Hitachi-Zosen
	Iroquois		15.24	14.60	222.50	15.93	Canadian-Vickers
1959	Setoishi	5	15.50	14.30	221.65	15.55	IHI
1957	Akiba	6	15.00	14.50	217.50	15.47	Mitsubishi
1973	Shimotokori	2	12.50	15.40	192.50	14.54	Hitachi-Zosen
1956	Sakuma	5	13.00	14.50	188.50	13.41	Mitsubishi
1970	Shinkawa	2	26.50	10.13	268.45	13.34	Kurimoto
1975	Shinano	3	30.00	9.10	273.00	12.19	IHI
1956	Ohuchihara	6	14.00	13.10	183.40	11.78	IHI
1970	Nakaura Suimon	10	32.00	8.45	270.40	11.21	IHI
1936	Quitzöbel	2	24.80	8.75	217.00	9.31	Thyssen-Klönne
1970	Kitakami I	1	50.00	6.10	305.00	9.13	Mitsubishi
1963	Agekawa	15	12.00	12.30	147.60	8.90	Hitachi-Zosen
1956	Yoake	8	15.00	11.00	165.00	8.90	IHI
1970	Niigata Ohzeki	5	41.20	6.40	263.68	8.28	IHI
1954	Pleidelsheim	1	33.30	6.90	229.77	7.78	DSD-NÖELL
1952	Pankei	4	18.20	9.30	169.26	7.72	IHI
1965	Nanairo	1	10.00	12.50	125.00	7.66	Kure
1973	Bow Creek	2	18.30	9.10	166.53	7.43	Newton-Chambers

Table 3.8 Spillway *Stoney* Gates

Year	Project	Quantity	Span (m)	Height (m)	Area (m²)	Hydrostatic load (MN)	Manufacturer
1933	Albruck-Dogern	5	25.00	12.00	300.00	17.66	
	Rivière-sur-Tarn	1	10.00	17.00	170.00	14.18	ALSTOM
1929	Cize-Bolozon	3	10.00	17.00	170.00	14.18	ALSTOM
1929	Pizançon	6	15.00	13.15	197.25	12.72	ALSTOM
	Chaton	3	30.50	8.36	254.98	10.46	
1925	Dnieper	25	24.00	9.00	216.00	9.54	
1932	Serrans	2	13.20	12.00	158.40	9.32	ALSTOM
1919	Beaumont-Monteux	3	18.77	10.00	187.70	9.21	ALSTOM
1936	Pickwick-Landing	24	12.20	12.20	148.84	8.91	
1951	Castel Giubil	4	20.00	9.50	190.00	8.85	Terni
1934	Noguera-Pallaresa	2	12.00	12.10	145.20	8.62	MAN
1919	Beaumont-Monteux	6	17.50	10.00	175.00	8.58	ALSTOM
1910	Tuilière	4	10.00	13.00	130.00	8.29	ALSTOM
1932	Safe Harbour	24	14.63	10.00	146.30	7.18	
	Alcalá del Rio	8	15.00	8.50	127.50	5.32	
	Sukkur	66	18.30	7.30	133.59	4.78	
1949	Cequeiros	3	15.00	8.00	120.00	4.71	
1917	Carpio	6	12.50	8.50	106.25	4.43	
1927	Sidi-Said-Machou	4	12.40	8.10	100.44	4.00	ALSTOM
1930	San Lorenzo	2	12.00	8.00	96.00	3.77	
1936	Rodriguez	9	9.14	9.14	83.54	3.75	MAN
1927	Ilha dos Pombos	5	12.00	7.60	91.2	3.40	R. Rapier

Table 3.9 Roller Gates

Year	Project	Quantity	Span (m)	Height (m)	Area (m²)	Hydrostatic load (MN)	Manufacturer
1938	Gallipolis	8	38.10	9.00	342.90	15.14	
1932	Marmet	4	30.50	7.92	241.56	9.38	
1934	Rock Island	9	30.40	7.92	240.77	9.35	
1919	Raanaasfos	1	45.00	6.50	292.50	9.33	MAN
1949	Koblenz	1	40.00	6.50	260.00	8.29	MAN
1933	Neckarzimmern	2	30.00	7.45	223.50	8.17	MAN
1935	Guttenbach	2	30.00	7.45	223.50	8.17	Krupp & others
1931	Hirschhorn	2	31.50	7.10	223.65	7.79	MAN
1919	Mörkfos-Solbergfos	3	20.00	8.75	175.00	7.51	MAN
1940	Koblenz	2	40.00	6.00	240.00	7.06	MAN
1936	Steinbach	2	30.00	6.90	207.00	7.01	MAN
1935	Rothenfels	2	30.00	6.75	202.50	6.70	Thyssen-Klönne
1932	Baldenay	3	33.50	6.35	212.73	6.63	Krupp & others
1917	Forshuvfud	2	17.00	8.85	150.45	6.53	MAN
1949	Offenbach	2	40.00	5.76	230.40	6.51	MAN
1935	Rockenau	2	30.00	6.50	195.00	6.22	Krupp & others
1934	Rock Island	2	30.48	6.42	195.68	6.16	
1938	Harbach	2	30.00	6.40	192.00	6.03	Thyssen-Klönne
1931	Neckarsteinach	2	33.00	6.00	198.00	5.83	Thyssen-Klönne
1933	Münster	2	23.30	7.00	163.10	5.60	MAN
1926	Hengstey	2	30.00	6.00	180.00	5.30	MAN
1928	Bell	2	35.00	5.48	191.80	5.16	
1910	Spokane	2	30.50	5.80	176.90	5.03	MAN
1926	Lewiston	3	32.00	5.64	180.48	4.99	
1910	Kibling	1	13.60	8.50	115.60	4.82	MAN

Table 3.10 Flap Gates

Year	Project	Quantity	Span (m)	Height (m)	Area (m²)	Hydrostatic load (MN)	Manufacturer
1965	St. Pantaleon	1	100.00	3.70	370.00	6.71	Waagner-Biro
1993	Bou Hanifia	3	25.00	6.75	168.75	5.59	Metalna
	Scrivener	5	30.00	6.00	180.00	5.30	
	Westhoe	2	45.40	4.77	216.56	5.07	DSD-NOELL
1934	Schafergrund	2	25.00	6.40	160.00	5.02	MAN
1949	Offenbach	1	40.00	5.00	200.00	4.91	MAN
1968	La Cave	3	30.50	5.50	167.75	4.53	MAN
1958	Ottendorf	2	30.00	5.50	165.00	4.45	MAN
1949	Isohaara	4	18.50	6.90	127.65	4.32	Tampella
1965	Couzon	4	35.00	4.80	168.00	3.96	ALSTOM
1960	Canberra	5	30.00	5.10	153.00	3.83	Thyssen-Klönne
1950	Skotofoss	2	24.50	5.40	132.30	3.50	Krupp
1972	Ruacana	5	56.00	3.50	196.00	3.36	ALSTOM
1947	Grossraming	2	22.50	5.50	123.75	3.34	VA TECH Hydro
1934	Heimbach	1	18.00	5.50	99.00	2.67	MAN
1960	Vichy	7	29.50	4.15	122.43	2.49	ALSTOM
	Polyphyton	3	12.90	6.00	77.40	2.28	
1950	Dona Aldonza	4	15.00	5.50	82.50	2.23	MAN
1932	Oued N'Fis	4	15.00	5.50	82.50	2.23	ALSTOM
1950	Pedro Marin	4	15.00	5.50	82.50	2.23	MAN
1966	Ragall	1	20.00	4.70	94.00	2.17	ZWAG
1952	Kisangani	3	13.00	5.60	72.80	2.00	MAN
1978	Sélingé	8	13.00	5.60	72.80	2.00	Krupp & others
1966	Feistritz	3	15.00	5.20	78.00	1.99	VA TECH Hydro
	Idaho Falls	2	45.00	3.00	135.00	1.99	ALSTOM
1969	Champagne	2	15.85	5.00	79.25	1.94	
1978	Breimsvatn	1	14.50	5.20	75.40	1.92	Kvaerner
1977	Bingsfoss	1	21.00	4.30	90.30	1.90	Kvaerner
1954	Kovada Ekridir	1	12.00	5.50	66.00	1.78	DSD-NÖELL
1968	Rendalen	1	20.00	4.20	84.00	1.73	Kvaerner
	Charnay	3	35.00	3.10	108.50	1.65	ALSTOM
1973	Pagny	3	36.50	3.00	109.50	1.61	ALSTOM
1957	Bärenburg	2	6.20	7.20	44.64	1.58	ZWAG

Table 3.11 Bear-Trap Gates

Year	Project	Quantity	Span (m)	Height (m)	Area (m²)	Hydrostatic load (MN)	Manufacturer
	Day	7	33.75	5.50	185.63	5.01	
	Flix		24.00	4.50	108.00	2.38	
1929	Guadalupe	3	25.00	4.00	100.00	1.96	
	Wiestal	1	32.00	3.10	99.20	1.51	
	Rheinau	4	26.00	3.30	85.80	1.39	MAN & others
	Montijo	3	20.00	3.50	70.00	1.20	
	Crevola d'Ossola	2	20.00	3.00	60.00	0.88	
1930	Tours-sur-Marne		34.00	1.97	66.98	0.65	
	Strubklamm	2	20.00	2.50	50.00	0.61	
1954	Schwellöd	1	19.10	2.50	47.75	0.59	Waagner-Biro
	Puentes de Princesa	4	17.40	2.50	43.50	0.53	
1956	Munchendorf	1	17.50	2.15	37.63	0.40	Waagner-Biro
1955	Nägeleinswehr	2	22.50	1.85	41.63	0.38	MAN
1929	Bestwig	2	15.00	1.90	28.50	0.27	Krupp
1928	Fröndenberg	2	15.50	1.50	23.25	0.17	Krupp

Table 3.12 Sector Gates

Year	Project	Quantity	Span (m)	Height (m)	Area (m²)	Hydrostatic load (MN)	Manufacturer
1924	Ilha dos Pombos (*)	3	45.00	7.40	333.00	12.09	R. Rapier
1960	Detzen	3	40.00	6.70	268.00	8.81	MAN
	Cruz del Eje (*)	2	27.00	8.00	216.00	8.48	
	San Lorenzo	2	36.00	6.50	234.00	7.46	
	Wintrich	3	40.00	5.90	236.00	6.83	Krupp
1963	Grevenmacher	2	40.00	5.90	236.00	6.83	Krupp
	Doiras (*)	2	28.00	7.00	196.00	6.73	
	Camarasa (*)	2	27.00	6.90	186.30	6.31	
1952	Upper Svirskaya	3	27.00	6.61	178.47	5.79	
1959	Enkirch	3	40.10	5.40	216.54	5.74	ALSTOM
1959	Muden	3	40.10	5.40	216.54	5.74	ALSTOM
1962	St. Aldegund	3	40.00	5.40	216.00	5.72	DSD-NOELL
1949	Hemelingen	2	54.00	4.60	248.40	5.60	Krupp
1948	Hadalsfallene	1	27.00	6.50	175.50	5.60	Kvaerner
1964	Breisach	4	45.00	5.00	225.00	5.52	Krupp
1959	Lehmen	3	40.00	5.30	212.00	5.51	Thyssen-Klönne
1909	Bremen	2	54.00	4.50	243.00	5.36	MAN
	Barasona	2	25.00	6.50	162.50	5.18	
1959	Geesthacht	4	50.00	4.55	227.50	5.08	Krupp & others
1940	Kettwig	2	43.70	4.80	209.76	4.94	Thyssen-Klönne
1952	Sarpsborg	1	28.00	6.00	168.00	4.94	Kvaerner
1919	Raanaasfos	2	50.00	4.00	200.00	3.92	MAN
	Biscarrues	4	18.50	6.50	120.25	3.83	
1960	Landesbergen	2	40.00	4.30	172.00	3.63	MAN
1952	Lower Svirskaya	1	31.00	4.70	145.70	3.36	
	Morris		21.20	5.50	116.60	3.15	
1963	Hunderfossen	1	20.00	5.50	110.00	2.97	Kvaerner

(*) Reinforced concrete structure

Table 3.13 Drum Gates

Year	Project	Quantity	Span (m)	Height (m)	Area (m²)	Hydrostatic load (MN)	Manufacturer
	Hamilton	1	91.00	8.50	773.50	32.25	
	Grand Coulee	11	41.00	8.50	348.50	14.53	
	Bhakra	2	41.00	8.50	348.50	14.53	
	Shasta	3	33.50	8.50	284.75	11.87	
1958	Liapootah	1	36.60	6.10	223.26	6.68	MAN
1941	Friant	3	30.50	5.50	167.75	4.53	
	Salinas		30.50	5.50	167.75	4.53	
1934	Hoover	8	30.40	4.90	148.96	3.58	
1951	Zvornik	8	18.00	6.30	113.40	3.50	Krupp
	Pitlochry	2	27.40	4.90	134.26	3.23	Glenfield
	Barnhart Island	4	22.90	5.18	118.62	3.01	MAN/Can. Vickers
	Norris	3	30.50	4.20	128.10	2.64	
1929	Myllykoski	1	18.00	4.90	88.20	2.12	Tampella
	Barnhart Island	2	15.20	5.18	78.74	2.00	MAN/Can. Vickers
1928	Easton Diversion Dam	1	19.50	4.42	86.19	1.87	
1921	Inkeroinen	1	20.00	4.00	80.00	1.57	Tampella
1931	Tampere	1	25.00	3.53	88.25	1.53	Tampella
1958	Schilsk	1	24.00	3.00	72.00	1.06	
1957	Sylvenstein	1	12.00	4.20	50.40	1.04	MAN

Table 3.14 Segment Gates with Flap

Year	Project	Quantity	Span (m)	Height (m)	Area (m²)	Hydrostatic load (MN)	Manufacturer
1968	Regua	1	26.00	15.73	408.98	31.56	Sorefame
1972	Valeira	1	26.00	15.73	408.98	31.56	Sorefame
1967	Carrapatelo	2	26.00	15.23	395.98	29.58	Sorefame
1966	Paldang	2	20.00	16.75	335.00	27.52	ALSTOM
1975	Altenwörth	2	24.00	15.00	360.00	26.49	Thyssen-Klönne
1977	Pocinho	1	26.00	14.13	367.38	25.46	Sorefame
1983	Greifenstein	6	24.00	14.50	348.00	24.75	VA TECH Hydro
1966	Carbonne	4	18.00	16.20	291.60	23.17	ALSTOM
1993	Freudenau	4	24.00	13.85	332.40	22.58	VA TECH Hydro
1960	Edling	3	15.00	17.00	255.00	21.26	VA TECH Hydro
1973	Ferlach	3	15.00	16.70	250.50	20.52	VA TECH Hydro
1980	Melk	6	24.00	13.00	312.00	19.89	VA TECH Hydro
1976	Song Loulou	7	14.00	17.00	238.00	19.85	Sorefame
1968	Weyer	3	18.00	14.80	266.40	19.34	VA TECH Hydro
1949	Donzère-Mondragon	1	45.00	9.15	411.75	18.48	ALSTOM
1978	Abwinden-Asten	1	24.00	12.50	300.00	18.39	Thyssen-Klönne
1976	Vaugris	4	21.00	12.8	268.80	16.88	ALSTOM
1965	Bourg-les-Valence	6	22.00	11.70	257.40	14.77	ALSTOM
1964	Villeneuve-sur-Lot	4	15.00	13.60	204.00	13.61	ALSTOM
1949	Donzère-Mondragon	4	31.50	9.15	288.23	12.94	ALSTOM
1963	St.Julien-Labrioulette	4	20.00	11.40	228.00	12.75	ALSTOM
1967	Azutan	1	20.00	11.30	226.00	12.53	ALSTOM
	Golfech	6	25.00	10.00	250.00	12.26	
1963	Gerstheim	3	20.00	10.30	206.00	10.41	ALSTOM

Table 3.15 Double-Leaf Fixed-Wheel Gates

Year	Project	Quantity	Span (m)	Height (m)	Area (m²)	Hydrostatic load (MN)	Manufacturer
1961	Aschach	7	24.00	16.00	384.00	30.14	Waagner-Biro
1975	Altenworth	2	24.00	15.70	376.80	29.02	Waagner-Biro
	Iron Gate	14	25.00	14.80	370.00	26.86	
1980	Crestuma	8	28.00	13.70	383.60	25.78	Sorefame
1955	Ybbs-Persenbeug	5	30.00	13.20	396.00	25.64	Thyssen-Klönne
	Montelimar	6	26.00	14.00	364.00	25.00	B. Seibert
1960	Beauchastel	6	26.00	13.50	351.00	23.24	ALSTOM
	Loriol	6	26.00	13.50	351.00	23.24	ALSTOM
1953	Rochemaure	6	26.00	13.50	351.00	23.24	ALSTOM
1958	Baix	6	26.00	13.50	351.00	23.24	ALSTOM
1963	Passau-Ingling	5	23.00	14.00	322.00	22.11	VA TECH Hydro
1943	Obernberg	5	23.00	13.50	310.50	20.56	Thyssen-Klönne
1953	Simbach-Braunau	5	23.00	13.50	310.50	20.56	Thyssen-Klönne
1959	Scharding	5	23.00	13.50	310.50	20.56	VA TECH Hydro
1967	Wallsee-Mitterkirchen	2	24.00	13.20	316.80	20.51	Waagner-Biro
1927	Ryburg-Schwörstadt	4	24.00	12.50	300.00	18.39	MAN
1966	Wallsee	6	24.00	12.50	300.00	18.39	VA TECH Hydro
	Marckolsheim	5	30.00	11.00	330.00	17.81	B. Seibert
1959	Losenstein	3	13.50	16.20	218.70	17.38	Waagner-Biro
1951	Sarobi	1	18.00	14.00	252.00	17.30	Krupp
1948	Flix	7	24.00	12.00	288.00	16.95	
1972	Otteensheim	5	24.00	12.00	288.00	16.95	VA TECH Hydro
	Belver	12	17.00	14.15	240.55	16.70	Sorefame

Table 3.16 Fixed-Wheel Gates with Flap

Year	Project	Quantity	Span (m)	Height (m)	Area (m²)	Hydrostatic load (MN)	Manufacturer
1966	Arzal	5	18.00	11.75	211.50	12.19	ALSTOM
1950	Tiszalök	3	37.00	8.00	296.00	11.62	Thyssen-Klönne
1958	Rosenheim	3	22.00	9.80	215.60	10.36	MAN
1960	Isola Serafini	11	30.00	8.00	240.00	9.42	Magrini-Galileo
1974	Kitagami II	2	50.00	6.10	305.00	9.13	Mitsubishi
1955	Langwedel	2	40.00	6.60	264.00	8.55	Krupp & others
1954	Schlüsselburg	2	40.00	6.50	260.00	8.29	Krupp & others
1933	Dörverden	2	41.60	6.35	264.16	8.23	Thyssen-Klönne
	Feldkirchen	4	15.00	10.30	154.50	7.81	DSD-NOELL
1937	Petershagen	3	40.00	6.25	250.00	7.66	Krupp & others
1955	Langwedel	1	30.00	6.60	198.00	6.41	Krupp & others
1948	Neuötting	5	18.00	8.50	153.00	6.38	MAN
	Poppenweiler	1	22.00	7.50	165.00	6.07	DSD-NOELL
1949	Niederaichbach	4	17.00	8.50	144.50	6.02	Krupp & others
1937	Harrbach	1	30.00	6.40	192.00	6.03	Krupp & others

Table 3.17 Outstanding Gates

Feature	Gate Type	Project	Span (m)	Height (m)	Area (m²)	Head on sill (m)	Load (MN)
Span (*)	Segment	New Waterway	360	22	7920	NA	740
Span	Fixed-wheel	Volga´s delta	110	12.93	1422	NA	NA
Height	Segment	Wuqiunangxi	19	23	437	23	49.3
Area	Drum	Hamilton	91	8.5	773.5	8.5	32.2
Head on sill	Slide	Beaver	2	3.18	6.3	285.9	17.7
Load on gate	Fixed-wheel	Tarbela	4.1	13.7	56.1	141	73.9

(*) Two gates are used to close the 360 m-wide canal
NA: not available

BIBLIOGRAPHY AND LIST OF GATE MANUFACTURERS

1. Gate Manufacturers

ALLIS CHALMERS	USA
ALSTOM	France, Brazil
ATB	Italy
BARDELLA	Brazil
COEMSA	Brazil
CONFAB	Brazil
DAVIESHIP	Canada
DEDINI	Brazil
ESCHER WYSS	Switzerland
HITACHI-ZOSEN	Japan
IMPSA	Argentina
ISHIBRAS	Brazil
ISHIKAWAJIMA HARIMA	Japan
J.M.VOITH	Germany
JESSOP	India
KRUPP	Germany
KURE	Japan

KURIMOTO	Japan
KVAERNER	Norway
M.A.N.	Germany
MAGRINI GALILEO	Italy
MITSUBISHI	Japan
NEPTUNE GLENFIELD	England
NEWTON CHAMBERS	England
DSD-NOELL	Germany
PACECO	USA
RIVA CALZONI	Italy
S.MORGAN SMITH	USA
SAKAI	Japan
SERMEC	Brazil
SOREFAME	Portugal
TAMPELLA	Finland
TERNI	Italy
THYSSEN KLÖNNE	Germany
TRIVENI STRUCTURALS	India
VA TECH Hydro	Austria
VICKERS CANADA	Canada
VOITH	Brazil
WAAGNER-BIRO	Austria
ZWAG	Switzerland

2.   Bibliography

Gomes Navarro, J.L. and Juan Aracil, J.: *Saltos de Agua y Presas de Embalse*, Madrid, Spain (1964).

Boissonnault, F.L.: Estimating Data for Reservoir Gates, *Transactions of ASCE*, Paper No. 2352 (1948).

Kollbrunner, C.F.: Hydraulic Steel Gates, *Proceedings of Research and Construction on Steel Engineering*, No. 13, Leeman Publishers, Zürich, Switzerland (Sept. 1950).

Bureau of Reclamation: *Valves, Gates and Steel Conduits*, Design Standards No. 7 (1956).

Davis, C.V. and Sorensen, K.E.: *Handbook of Applied Hydraulics*, McGraw Hill, 3rd Edition (1969).

REFERENCES

1. Csallner, K.: Strömungstechnisch und Konstruktive Kriterien für die Wahl Zwischen Druck und Zugsegment als Wehrverschluss, *Versuchsanstalt fur Wasserbau der Technischen Universität München*, Report No. 37 (1978).

2. Buzzell, Dow A.: Trends in Hydraulic Gate Design, *Transactions of ASCE*, Paper No. 2908 (1957).

3. Daumy, G.: Aperçu sur l'Évolution de la Technique des Vannes pour Installations Hydroélectriques, *La Houille Blanche*, July/August (1962).

4. Thomas, Henry H.: *The Engineering of Large Dams*, John Wiley and Sons (1976).

Author's note: Part of this chapter was first published by the author in the International Water Power and Dam magazine on April 1981, under the title "Hydraulic Gates – The state-of-the-Art". It is reproduced with the Editor's agreement.

Chapter 4

Hydrostatics

4.1 INTRODUCTION

When dimensioning a gate, the first step is to calculate the water thrust acting on the skin plate for the various gate-opening positions. Its maximum value occurs with the gate closed and subjected to the maximum head water level.

For gates with water on both sides of the skin plate, the maximum water thrust corresponds to the most unfavorable unbalanced level between the upstream and the downstream reservoirs.

4.2 VERTICAL LIFT GATES

4.2.1 WEIR GATES

For gates with only one side of the skin plate in contact with water, the maximum water thrust is given by the following formula:

$$W = \tfrac{1}{2}\,\gamma\,B\,H^2 \tag{4.1}$$

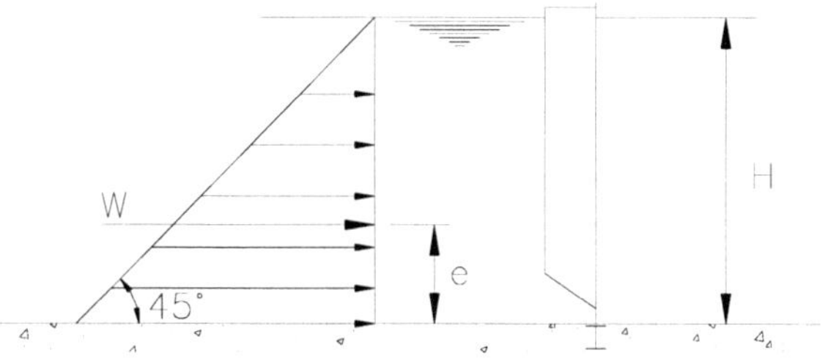

Fig. 4.1 Pressure diagram on flat weir gates with water on one side

where:
γ = specific weight of water = 9.81 kN/m^3
B = span of side seals
H = maximum headwater on sill.

Its line of action is normal to the skin plate and passes through the center of pressure of the surface, that is, at a distance above the sill of

$$e = \frac{1}{3} H \qquad\qquad (4.2)$$

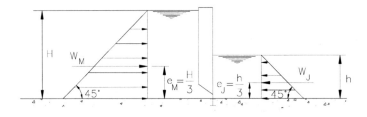

Fig. 4.2 Pressure diagram on flat weir gates with water on both sides

For the gate of Figure 4.2, with both sides of the skin plate in contact with water, the water thrust due to the upstream reservoir is:

$$W_M = \tfrac{1}{2}\,\gamma\,B\,H^2$$

and its centroid is

$$e_M = 1/3\ H$$

On the downstream side,

$$W_J = \tfrac{1}{2}\,\gamma\,B\,h^2 \qquad\qquad (4.1a)$$

and

$$e_J = 1/3\ h \qquad\qquad (4.2a)$$

where h is the minimum downstream headwater on the sill.

The resultant water thrust will then be,

$$W = W_M - W_J = \tfrac{1}{2}\,\gamma\,B\,(H^2 - h^2) \qquad\qquad (4.3)$$

Considering z as the vertical distance between the centroid of W and the sill and taking moments about the sill,

$$W_M\, e_M - W_J\, e_J = W\, z$$

$$\therefore z = (W_M\, e_M - W_J\, e_J)/W \quad = \frac{\left(\frac{1}{2}\gamma\, BH^2\right)\left(\frac{1}{3}H\right) - \left(\frac{1}{2}\gamma\, Bh^2\right)\left(\frac{1}{3}h\right)}{\frac{1}{2}\gamma\, B\left(H^2 - h^2\right)}$$

Finally,

$$z = \frac{H^3 - h^3}{3(h^2 - h^2)} \tag{4.4}$$

4.2.2 SUBMERGED GATES

For submerged gates with only one side of the skin plate in contact with water, the water thrust is calculated by

$$W = \gamma\, B\, h\, (H - h/2) \tag{4.5}$$

where

γ = specific weight of water = 9.81 kN/m^3
B = span of side seals
H = maximum headwater on sill
h = gate sealing height.

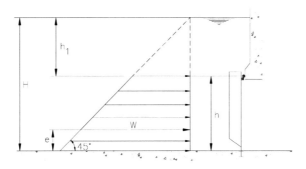

Fig. 4.3 Pressure diagram on flat submerged gates with water on one side

The pressure diagram has a trapezoidal shape. In this case, the position of the resultant water thrust is given by

$$e = \frac{h}{3}\left(1 + \frac{H - h}{2H - h}\right) \tag{4.6}$$

Substituting in the above equations,

$$h = H - h_1 \qquad\qquad (4.7)$$

results, after simplification,

$$W = \tfrac{1}{2}\,\gamma\,B\,(H^2 - h_1^2) \qquad\qquad (4.5a)$$

and

$$e = \left(\frac{H - h_1}{3}\right)\left(\frac{H + 2h_1}{H + h_1}\right) \qquad\qquad (4.6\,a)$$

where h_1 is the vertical distance between the free surface of water and the top seal.

For submerged gates, with both sides of the skin plate in contact with water, the resultant water thrust is the difference between the hydrostatic forces due to each reservoir.

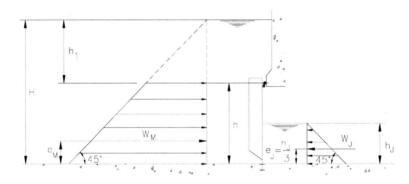

Fig. 4.4 Pressure diagram on flat submerged gates with water on both sides

. Upstream water thrust

$$W_M = \tfrac{1}{2}\,\gamma\,B\,(H^2 - h_1^2)$$

and

$$e_M = \left(\frac{H - h_1}{3}\right)\left(\frac{H + 2h_1}{H + h_1}\right)$$

. Downstream water thrust

$$W_J = \tfrac{1}{2}\,\gamma\,B\,h_J^2$$

where h_J is the minimum depth of the tailwater.

The centroid of W_J is calculated by

$$e_J = 1/3\ h_J$$

since the pressure diagram has a triangular shape.

The resultant water thrust is:

$$W = W_M - W_J = \tfrac{1}{2} \gamma B (H^2 - h_1^2) - \tfrac{1}{2} \gamma B h_J^2 = \tfrac{1}{2} \gamma B (H^2 - h_1^2 - h_J^2) \quad (4.8)$$

Considering z as the vertical distance between the line of action of W and the gate sill and taking moments about the sill, results in

$$W_M e_M - W_J e_J = W z$$

$$\therefore z = (W_M e_M - W_J e_J)/W$$

After the necessary substitutions and simplification,

$$z = \frac{(H - h_1)^2 (H + 2h_1) - h_j^3}{3(H^2 - h_1^2 - h_j^2)} \qquad (4.9)$$

4.2.3 SPACING OF HORIZONTAL BEAMS

From the viewpoint of costs, it is worthwhile to have all horizontal beams equally loaded, in order to design a unique cross section for all beams. This is achieved by dividing the pressure diagram into equivalent areas and locating the centerline of each beam in the centroid of each area.

a) Weir gates

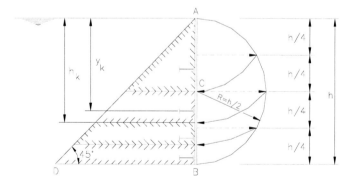

Fig. 4.5 Flat weir gates – Division of the pressure diagram in four equivalent areas

The graphic method for division of the pressure diagram into equivalent areas is shown in Figure 4.5. The sequence is:
- to divide the height h (segment AB) in n equal parts;
- to draw a semi-circumference centered on C;

- to horizontally link the divisions of the segment AB to the semi-circumference;
- centered on A, transport to the segment AB the intersection points marked in the semi-circumference.

The new points marked on the segment AB outline, in the load triangle ABD, n surfaces of equivalent areas. The horizontal beams should be located in the centroid of each area; it should be noted that all areas have a trapezoidal shape, except the top one, which is triangular. The outlining of the equivalent areas and the positioning of the horizontal beams can be made, alternatively, by means of an analytical procedure, with the use of the following equations:

- depth h_k

$$h_k = h \sqrt{\frac{k}{n}} \qquad \text{(where } k = 1, 2, 3 ..., n)} \qquad (4.10)$$

- depth of the horizontal beams

$$y_k = \frac{2h}{3\sqrt{n}}\left[k^{3/2} - (k-1)^{3/2} \right] \qquad (4.11)$$

where n is the quantity of beams or areas.

b) Submerged gates

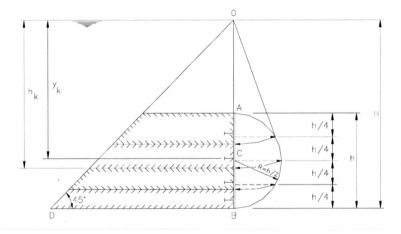

Fig. 4.6 Flat submerged gates – Division of the pressure diagram in four equivalent areas

The graphic method for division of the load diagram into n equivalent areas is similar to that suggested for the weir gates, except that the compass should be centered on point O located on the free surface of water, in the extension of the segment AB.

- Depth h_k

$$h_k = H\sqrt{\frac{k+\beta}{n+\beta}} \qquad\qquad \text{(where } k = 1, 2, 3, ..., n\text{)} \qquad\qquad (4.12)$$

where

$$\beta = \frac{n(H-h)^2}{H^2 - (H-h)^2} \qquad\qquad (4.13)$$

H = maximum headwater on sill
h = gate sealing height
n = quantity of areas (or beams)

- position of horizontal beams

$$y_k = \frac{2H}{3\sqrt{n+\beta}}\left[(k+\beta)^{3/2} - (k-1+\beta)^{3/2}\right] \qquad\qquad (4.14)$$

With this method, all beams will be equally loaded. It can be also used to determine the position of equally loaded wheels of fixed-wheel gates.

Example 4.1 A vertical lift gate 4 m wide by 6 m high has six equally loaded horizontal beams and is subjected to a headwater on the sill of 10 m. Determine the maximum water thrust on the gate and its position. Define the location of the horizontal beams.

Solution:

B = gate span = 4 m
h = gate seal height = 6 m
H = max. headwater on sill = 10 m
γ = specific weight of water = 9.81 kN/m³.

Equation 4.5 gives the maximum water thrust:

W = 9.81 · 4 · 6 (10 - 6/2) = 1648 kN

The vertical distance between W and the sill is (Eq. 4.6):

$$e = \frac{6}{3}\left(1 + \frac{10-6}{2\cdot10-6}\right) = 2.57\,\text{m}$$

Depth h_k :

Equation 4.13 gives:

$$\beta = \frac{6(10\text{-}6)^2}{10^2 \text{-}(10\text{-}6)^2} = 1.1428$$

The depth h_k will be (Eq. 4.12):

$$h_k = 10\sqrt{\frac{k+1.1428}{6+1.1428}} = 3.742\sqrt{k+1.1428}$$

and

$$h_1 = 3.742\,(1 + 1.1428)^{1/2} = 5.48 \text{ m}$$
$$h_2 = 3.742\,(2 + 1.1428)^{1/2} = 6.63 \text{ m}$$
$$h_3 = 3.742\,(3 + 1.1428)^{1/2} = 7.62 \text{ m}$$
$$h_4 = 3.742\,(4 + 1.1428)^{1/2} = 8.49 \text{ m}$$
$$h_5 = 3.742\,(5 + 1.1428)^{1/2} = 9.27 \text{ m}$$
$$h_6 = 3.742\,(6 + 1.1428)^{1/2} = 10 \text{ m}.$$

Equation 4.14 gives the depth of the horizontal beam axis:

$$y_k = \frac{2 \times 10}{3\sqrt{6+1.1428}}\left[(k+1.1428)^{3/2} - (k-1+1.1428)^{3/2}\right] =$$
$$= 2.494\left[(k+1.1428)^{3/2} - (k+0.1428)^{3/2}\right]$$

So,

$$y_1 = 2.494\left[(1+1.1428)^{3/2} - (1+0.1428)^{3/2}\right] = 4.78 \text{ m}$$

and, successively,

$$y_2 = 6.07 \text{ m}$$
$$y_3 = 7.14 \text{ m}$$
$$y_4 = 8.06 \text{ m}$$
$$y_5 = 8.89 \text{ m}$$
$$y_6 = 9.64 \text{ m}.$$

4.3   RADIAL GATES

The line of action of the water thrust on radial gates passes through the center of curvature of the skin plate. For ease of calculation the resultant water thrust is determined from its horizontal and vertical components.

The formulae presented below are generic and can be indistinctly used for weir and submerged segment gates, sector gates and flap gates with radial skin plate.

The horizontal component of the water thrust is:

$$W = \gamma\, B\, h\, (H - h/2) \tag{4.15}$$

The vertical component is:

$$W_v = \gamma BR[D_m(\cos \alpha_s - \cos \alpha_i) + R(\alpha_i - \alpha_s)/2 + R(\sin \alpha_s \cos \alpha_s - \sin \alpha_i \cos \alpha_i)/2] \tag{4.16}$$

where

γ = specific weight of water = 9.81 kN/m^3
B = side seal span
R = skin plate radius (measured on the wet surface)
H = maximum headwater on sill
h = gate sealing height
D_m = difference between the elevations of the water level and the center of curvature of the skin plate

$$\alpha_s = \text{arc sin } D_s/R \tag{4.17}$$
$$\alpha_i = \text{arc sin } D_i/R \tag{4.18}$$

D_s = difference between the elevations of the center of curvature of the skin plate and the top seal (for submerged gates) or the water level (in case of weir gates)
D_i = difference between the elevations of the center of curvature of the skin plate and the sill.

NOTES:

a) D_m, D_s and D_i refer to difference of elevations and may assume positive or negative sign, depending on the gate arrangement;
b) The angles α_s and α_i are taken in radians and can also assume a positive or negative sign;
c) The direction of the vertical component is given by its sign:
   - positive: upward
   - negative: downward.

The magnitude and direction of the resultant water thrust are calculated by:

$$W = \sqrt{W_h{}^2 + W_v{}^2} \tag{4.19}$$

$$\beta = \text{arc tan } W_v\, /\, W_h \tag{4.20}$$

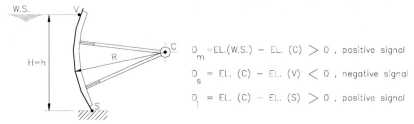

(a) Weir segment gate with center of curvature of the skinplate below the water level.

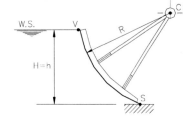

(b) Weir segment gate with center of curvature of the skinplate above the water level.

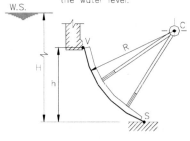

(c) Submerged segment gate with center of curvature of the skinplate above the top seal.

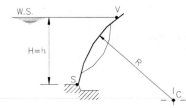

(d) Flap gate with center of curvature of the skinplate below the hinge axis.

Fig. 4.7 Radial gates – Parameters for calculation of the maximum hydraulic thrust

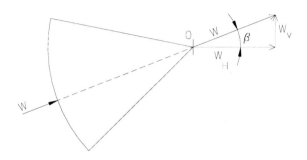

Fig. 4.8 Direction of the maximum hydraulic thrust on radial gates

Example 4.2 Weir segment gate

Determine the total water thrust on the segment gate of the Figure 4.9, with a width of 15 m and radius of 19 m.

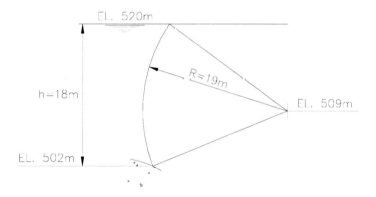

Fig. 4.9 Spillway segment gate

Solution:

For weir gates, the maximum headwater on the sill is equal to the gate sealing height, that is, $H = h$. In this case,

$H = h = $ El. 520 - El. 502 = 18 m

Also,
  B = side seal span = 15 m
  R = skin plate radius = 19 m

D_m = El. 520 - El. 509 = 11m
D_s = El. 509 - El. 520 = - 11m
D_i = El. 509 - El. 502 = 7 m

Therefore,
α_s = arc sin (-11/19) = -35.377 degrees = - 0.6174 rad
α_i = arc sin 7/19 = 21.618 degrees = 0.3773 rad
$\cos \alpha_s$ = 0.8154 $\sin \alpha_s$ = - 0.5789
$\cos \alpha_i$ = 0.9297 $\sin \alpha_i$ = 0.3684

The horizontal component is (Eq. 4.15):

W_h = 9.81·15·18(18 - 18/2) = 23838 kN

Equation 4.16 determines the vertical component:

W_v = 9.81·15·19{11(0.8154-0.9297)+1/2·19[0.3773-(-0.6174)]+
          +1/2·19[(-0.5789)·0.8154-0.3684·0.9297]} = 1269.6 kN

Magnitude of W (Eq. 4.19):

$$W = \sqrt{23838^2 + 1269.6^2} = 23872 \ \text{kN}$$

Direction of W (Eq. 4.20):

β = arc tan 1269.6/23838 = 3.049 degrees

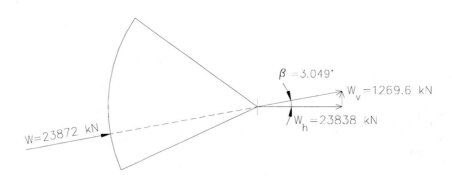

Fig. 4.10 Radial gates - Direction of the hydraulic thrust (Example 4.2)

Example 4.3 Submerged segment gate
Determine the magnitude and the direction of the maximum water thrust on the gate of
Figure 4.11, with a skin plate radius of 6.5 m and a span of 2.5 m.

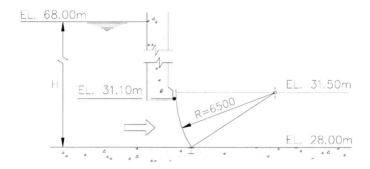

Fig. 4.11 High-pressure segment gate

Solution:

In this case,

B = 2.5 m
R = 6.5 m
D_m = El. 68 - El. 31.5 = 36.5 m
H = El. 68 - El. 28 = 40 m
h = El. 31.1 - El.28 = 3.1 m
D_s = El. 31.5 - El. 31.1 = 0.4 m
D_i = El. 31.5 - El. 28 = 3.5 m
α_s = arc sin 0.4/6.5 = 3.5281 degrees = 0.0616 rad
α_i = arc sin 3.5/6.5 = 32.5790 degrees = 0.5686 rad
$\cos \alpha_s$ = 0.9981 $\sin \alpha_s$ = 0.0615
$\cos \alpha_i$ = 0.8427 $\sin \alpha_i$ = 0.5385

Horizontal component of W (Eq. 4.15):

W_h = 9.81 · 2.5 · 3.1(40 - 3.1/2) = 2923.2 kN

Vertical component of W (Eq. 4.16):

W_v = 9.81·2.5·6.5[36.5(0.9981-0.8472)+1/2·6.5(0.5686-0.0616) +
        +1/2·6.5(0.0615·0.9981-0.5385·0.8427)] = 964 kN

Magnitude of W (Eq. 4.19):

$$W = \sqrt{2923.2^2 + 964^2} = 3078 \text{ kN}$$

Direction of W (Eq. 4.20):

β = arc tan 964/2923.2 = 18.251 degrees

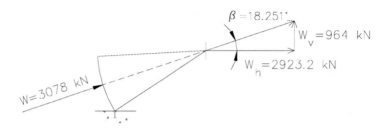

Fig. 4.12 Direction of W (Example 4.3)

Example 4.4 Flap gate with radial skin plate
Determine the magnitude and the direction of the maximum water thrust on the flap gate of Figure 4.13 with a 12 m span and a skin plate radius of 6 m.

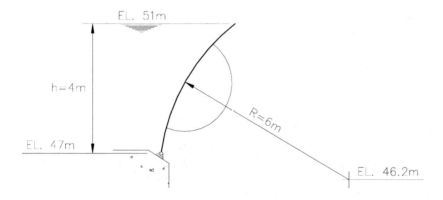

Fig. 4.13 Flap gate

Solution:
B = 12 m
R = 6 m
D_m = El. 51 - El. 46.2 = 4.8 m
$H = h$ = El. 51 - El. 47 = 4 m
D_s = El. 46.2 - El. 51 = -4.8 m
D_i = El. 46.2 - El. 47 = -0.8 m
α_s = arc sin (-4.8/6) = - 53.130 degrees = - 0.9273 rad
α_i = arc sin (-0.8/6) = - 7.6623 degrees = - 0.1337 rad

$\cos \alpha_s = 0.6000$ $\sin \alpha_s = -0.8000$
$\cos \alpha_i = 0.9911$ $\sin \alpha_i = -0.1333$

Horizontal component of W (Eq. 4.15):

$W_h = 9.81 \cdot 12 \cdot 4(4 - 4/2) = 941.8$ kN

Vertical component of W (Eq. 4.16):

$W_v = 9.81 \cdot 12 \cdot 6\{4.8(0.6000 - 0.9911) + 1/2 \cdot 6[-0.1337 - (-0.9273)] +$
     $+ 1/2 \cdot 6[(-0.8000)0.6000 - (-0.1333)0.9911]\} = -381.4$ kN

Magnitude of W (Eq. 4.19):

$$W = \sqrt{941.8^2 + (-381.4)^2} = 1016 \text{ kN}$$

Direction of W (Eq. 4.20):

$\beta = $ arc tan $(-381.4/941.8) = -22.05$ degrees

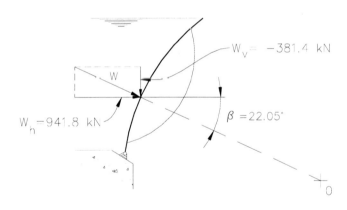

Figure 4.14 Flap gate – Direction of W (Example 4.4)

Chapter 5

Structural Design

5.1 LOAD CASES

The Brazilian standard NBR-8883 establishes three load cases for design of gates, according to the frequency of occurrence and the nature of the loads and the probability of their coincidence:

- Normal load case - considers the most unfavorable values and combinations of the hydrostatic loads at normal water levels (including the influence of waves), hydrodynamic effects, friction forces, dead weight, buoyancy, transit loads and driving forces.
  The simultaneous occurrence of these loads and water levels, as well their combinations, should only be considered when possible and probable.
- Occasional load case - considers the loads which occur less frequently, such as:
  . hydrostatic and hydrodynamic forces at unusual water levels;
  . wind loads;
  . temperature effects;
  . friction by ships;
  . ice impact and pressure.

  The simultaneous occurrence of these loads and their combinations with each other and with those of the preceding item should only be considered when possible and probable.
- Exceptional load case - considers loads occurring during transportation, erection, maintenance services or other exceptional cases, as well as the following:
  . hydrodynamic effects and overloads due to the driving forces in the event of lining or penstock failure;
  . asymmetrical loads and overloads due to the driving forces caused by jamming of the gate by foreign bodies or damage to bearings, rollers or hinges;
  . impact of ships;
  . seismic effects;
  . changes in conditions of support.

The simultaneous occurrence of these loads and those of the items *a* and *b* should only be considered when their combination is possible and probable.

Hydrodynamic forces are generally determined by means of model tests. These tests are made in hydraulic laboratories and may substantially increase the cost of small gates. In these cases, it is usual to determine the hydrodynamic forces by analytical methods, provided a previous agreement is reached between the customer and the gate supplier.

The influence of the operating forces on the structural elements should be considered according to the hoist rated capacity for the normal load case, or with the maximum capacity, for the exceptional case. The maximum capacity to be considered, in the case of fluid power systems, is the one corresponding to the relief valve pressure; in the case of mechanical hoists, that of the limiting device or, in its absence, that of the maximum torque of the driving motor. In gravity-closure gates, the preponderance of the closing forces should be proved with a safety margin of 20 per cent, for the normal load case, and 15 per cent, for the exceptional case.

The wind direction should be assumed to act horizontally. The surface exposed to the wind should be that resulting from parallel projection on a plane lying vertical to the direction of the wind. The wind pressure, provided more unfavorable conditions are not prescribed, is considered as:
- for gates in motion, equal to 500 N/m^2;
- for gates at rest, equal to 1000 N/m^2.

The thermal effects are investigated separately for temperature fluctuation in relation to the erection temperature and for uneven action on the structure. If more unfavorable conditions are not prescribed, the following temperature fluctuations should be considered:
- for gates lifted out of water temporarily in their entirety or for a major part of the time: $\pm$ 30 degrees C;
- for gates submerged for a major part of the time or lifted out of water in an environment protected from large temperature fluctuations: $\pm$ 20 degrees C;
- for gates completely submerged: $\pm$ 10 degrees C.

Loads due to waves are considered according to local conditions. In lock gates, an increment of 0.25 m to the water level should be considered, unless more unfavorable conditions are prescribed.

The impact force of ships against lock gates is calculated according to the mass of the greatest ship totally loaded anticipated for the waterway. If more unfavorable conditions are not prescribed, the gates should resist a force whose numerical value, expressed in kN, is equal to twice the numerical value of the ship mass, expressed in 10^6 kg. Loads resulting from ship friction with lock gates in the open position should be considered in the dimensioning. Provided local conditions do not impose different values, a horizontal force of 50 kN should be considered applied within the range of the water line in the direction of travel and, simultaneously, a horizontal force of 100 kN, acting perpendicularly to the direction of travel.

Loads resulting from changes in the support conditions due to differential settlement of the foundation or to concrete structure displacements should be taken into account in the gate design, according to the prevailing local conditions.

In gates designed to serve as a bridge for passage of vehicles, traffic loads should be considered according to the applicable standards.

The seismic influence is taken into account in the gate design with their effect simulated as a horizontal force of magnitude equal to the gate mass multiplied by the probable horizontal seismic acceleration in the region. The possible occurrence of the resonance phenomena and its effects should be investigated.

The influence of ice pressure and impact is considered according to the local conditions.

5.2 ALLOWABLE STRESSES

The allowable stresses are determined according to the yield strength of the material and should take into account the load case.

For structural elements, the allowable stresses correspond to the yield strength multiplied by the coefficients of Table 5.1.

Table 5.1 Coefficients for Allowable Stresses, Structural Elements

Type of Stress	Load Case		
	Normal	Occasional	Exceptional
Tension and bending stress	0.68	0.76	0.89
Bending stress if a stability proof is required	0.59	0.68	0.79
Shear	0.39	0.44	0.51
Combined stress	0.76	0.82	0.92
Combined stress in the skin plate	0.87	0.87	0.92

The allowable bearing pressure between pins and plates in structural elements can be determined by applying the following reduction coefficients to the yield strength:
. normal load case,   0.8
. occasional load case,   0.9
. exceptional load case,   0.9

For bi-axial stress conditions, the comparison stress is given by the formula:

$$\sigma^* = \sqrt{\sigma_x^2 + \sigma_y^2 - \sigma_x \sigma_y + 3\tau_{xy}^2} \qquad (5.1)$$

where:
σ_x = sum of stresses along x-axis
σ_y = sum of stresses along y-axis
τ_{xy} = shear stress in the plane normal to x-axis or y-axis.

In February 1996 the Brazilian standard NBR-8883 *Calculation of Hydraulic Gates* was revised and updated. The most important modification occurred with the allowable stresses for structural elements. In its 1985 original version, limits were established for the resultant stresses in the main axis, in parallel with a check of the combined stresses. However, according to the Hencky-von Mises theory, in a multiple state of stresses it is sufficient to check the combined stresses at the point, even in the case where the resultant stresses in each main axis reach high values. Based on this concept, the limits for the axial resultant stress have been eliminated.

In the most recent version of the document *Technical Standards for Gates and Penstocks,* published by the Japanese Hydraulic Gate and Penstock Association, the allowable stresses in the main axis are limited and a check of the combined stress in two or more axial directions should be provided. This requirement, adopted in the original version of the Brazilian standard NBR-8883 and throughout this book, is conservative and results in heavier and safer equipment. For structural design, the Japanese standards establish three groups of gates:
-   gate in full-time use;
-   high-pressure gate in full-time use;
-   gate in part-time use.

According to the structural steel used in the fabrication and using the above classification, different levels of axial and combined allowable stresses are established.

In May 1998 a new version of the DIN-19704 German standard (*Stahlwasserbauten Berechnungsgrundlagen*) was published. The most substantial changes were:
-   calculation of the steel structure is based on the concept of limit state design using partial safety factors and combination factors; forces are classified as permanent, variable (frequent and rare), extreme or accidental;
-   steel structures and machinery elements are not regarded as being constructions under static load. Therefore, fatigue strength proofs are required;
-   a new concept is applied for calculating the resistance of plain bearings. The basis of calculation for maintenance-free bearings is now the annual sliding distance;
-   for steel structures a service life of 70 years has to be assumed; for machinery elements and electrical equipment, 35 years. This does not refer to wear parts;
-   manufacturing tolerances are given for the steel structures;
-   hydraulic drives are dealt with in detail;
-   new regulations are given for calculating the resistance of wheels and rails.

For mechanical elements, the allowable stresses correspond to the yield limit multiplied by the coefficients of Table 5.2. In general, the following gate components are designed as mechanical elements: wheels, shafts, bearings and bushings, lifting beam hooks, chains, gears, lifting pins, screws, dogging devices, mechanical and hydraulic hoists.

Table 5.2 Coefficients for Allowable Stresses, Mechanical Elements

Type of Stress	Load Case		
	Normal	Occasional	Exceptional
Tension and compression	0.40	0.50	0.80
Shear	0.23	0.29	0.46

The comparison stress, for mechanical elements subjected to a normal stress and a shear stress, is given by

$$\sigma^* = \sqrt{\sigma^2 + 3\tau^2}$$
(5.2)

where:

  σ = tensile or compressive stress

  τ = shear stress.

The comparison stress should not exceed 1.25 times the allowable stress determined through Table 5.2 for the cases of normal and occasional loads and, the yield strength, in the case of exceptional loads.

  Under special circumstances (safety increase, for example) and at the owner's discretion, allowable stresses below those determined according to Tables 5.1 and 5.2 may be specified.

5.3 SKIN PLATE

5.3.1 THICKNESS

The skin plate comprises the greatest part of the gate weight. Therefore, the designer should pay particular attention to its dimensioning in order to achieve the smallest possible thickness consistent with the required structural strength. Generally, the smallest thickness used in skin plates is 8 mm, which permits welding of the reinforcing elements without significant plate warping. For small weir gates receiving good maintenance services, 6.5 mm thick plates may be used, provided precautions against distortion are taken during welding. On the other hand, thick plates up to 40 mm are used in high-head gates, as in the case of the fixed-wheel gates used for closure of the Itaipu diversion sluices, with 6.5 m span, 22 m high, subjected to a water head of 140 m on the sill. The plate thickness is determined through comparative studies of cost between the various spacing alternatives of stiffeners and beams welded therein.

5.3.2 PLATE STRESSES

According to the NBR-8883 standard, the plate bending stresses from water pressure are calculated with the theory of plates based on the theory of the elasticity, through the formula

$$\sigma = \pm \frac{k}{100} p \frac{a^2}{t^2} \qquad (5.3)$$

where:

  k = non-dimensional factor obtainable from Table 5.3, in function of the ratio b/a (support length of the modules formed by the beams and/or stiffeners) and the support conditions of the module

  p = water pressure relative to the module center

  a = minor support length

  t = plate thickness.

Table 5.3 k-Coefficient (NBR-8883)

b/a	Unfixed mounting of the 4 edges		Rigid fixing of the 4 edges				Rigid fixing of 3 edges and unfixed mount of the fourth edge							
	(a)		(b)				(c)				(d)			
	$\pm\sigma_{1x}$	$\pm\sigma_{1y}$	$\pm\sigma_{1x}$	$\pm\sigma_{1y}$	$\pm\sigma_{4y}$	$\pm\sigma_{3x}$	$\pm\sigma_{1x}$	$\pm\sigma_{1y}$	$\pm\sigma_{4y}$	$\pm\sigma_{3x}$	$\pm\sigma_{1x}$	$\pm\sigma_{1y}$	$\pm\sigma_{2y}$	$\pm\sigma_{3x}$
∞	75.0	22.5	25.0	7.5	34.2	50.0	37.5	11.3	47.2	75.0	25.0	7.5	34.2	50.0
3.00	71.3	24.4	25.0	7.5	34.3	50.0	37.4	12.0	47.1	74.0	25.0	7.6	34.2	50.0
2.50	67.7	25.8	25.0	8.0	34.3	50.0	36.6	13.3	47.0	73.2	25.0	8.0	34.2	50.0
2.00	61.0	27.8	24.7	9.5	34.3	49.9	33.8	15.5	47.0	68.3	25.0	9.0	34.2	50.0
1.75	55.8	28.9	23.9	10.8	34.3	48.4	30.8	16.5	46.5	63.2	24.6	10.1	34.1	48.9
1.50	48.7	29.9	22.1	12.2	34.3	45.5	27.1	18.1	45.5	56.5	23.2	11.4	34.1	47.3
1.25	39.6	30.1	18.8	13.5	33.9	40.3	21.4	18.4	42.5	47.2	20.8	12.9	34.1	44.8
1.00	28.7	28.7	13.7	13.7	30.9	30.9	14.4	16.6	36.0	32.8	16.6	14.2	32.8	36.0

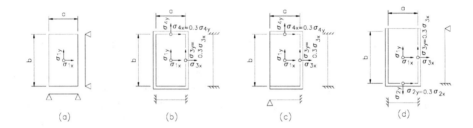

The maximum deflection occurs at the center of the plate, and is given by:

$$f = \frac{\alpha \; p \; a^4}{E \; t^3}$$

(5.4)

where

E = modulus of elasticity
α = coefficient given in Table 5.4, as a function of the plate dimensions.

Table 5.4 α-Coefficient (Timoshenko)

b/a	1.0	1.2	1.4	1.6	1.8	2.0	∞
α	0.0138	0.0188	0.0226	0.0251	0.0267	0.0277	0.0284

Spacing of horizontal beams is influenced by the skin plate thickness: the thicker the plate, the more spaced the beams may be. If, for constructive or economic reasons, the design specifies equal sized horizontal beams, their spacing should be gradual, increasing with height, so that all be subjected to the same part of the water thrust. If the beams are equally spaced, the lower ones must be stronger since they are subjected to greater loads.

Example 5.1 Find the skin plate bending stresses for the module indicated below, located at a depth of 36 m. The plate thickness is 19 mm. Find, also, the deflection at the center of the plate.

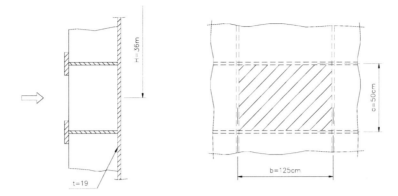

Fig. 5.1

Solution:

The ratio of the support lengths is

$b/a = 125/50 = 2.5$

Table 5.3 gives, for $b/a = 2.5$ and rigid fixing of four edges, the following coefficients:

$k_{1x} = 25$
$k_{1y} = 8$
$k_{4y} = 34.3$
$k_{3x} = 50$

Since the module center is at a depth of 36 m, the corresponding hydrostatic pressure will be:

$P = \gamma H = 9.81 0^{-6} \, N/mm^2 \cdot 6000 \, mm = 0.353 \, N/mm^2 = 0.353 \, MPa$

Therefore, according to Equation 5.3, we have:

$\sigma_{1x} = \pm 0.353 \cdot 25/100(500/19)^2 = \pm 61.11 \, MPa$
$\sigma_{1y} = \pm 0.353 \cdot 8/100(500/19)^2 = \pm 19.56 \, MPa$
$\sigma_{4y} = \pm 0.353 \cdot 34.3/100(500/19)^2 = \pm 83.85 \, MPa$
$\sigma_{3x} = \pm 0.353 \cdot 50/100(500/19)^2 = \pm 122.23 \, MPa.$

Still, by Table 5.3,

$\sigma_{4x} = \pm 0.3 \, \sigma_{4y} = \pm 0.3 \cdot 83.85 = 25.15 \, MPa$
$\sigma_{3y} = \pm 0.3 \, \sigma_{3x} = \pm 0.3 \cdot 122.23 = 36.67 \, MPa.$

The shape of the deflected skin plate determines the stress type (tension or compression). In the present case, the skin plate deflects due to the water pressure as shown in Figure 5.2.

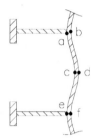

Fig. 5.2 Schematic arrangement of the deflected skin plate

At points a, d and e, the fibers are stretched and, by convention, the stresses are considered positive. At points b, c and f, the fibers are compressed (therefore the stresses are, by convention, negatives). As result, the plate stresses in the downstream face of the skin plate are as shown in Figure 5.3.

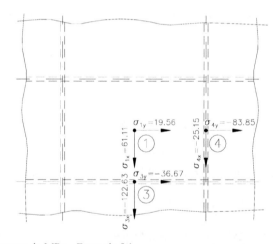

Fig. 5.3 Skin plate stresses, in MPa – Example 5.1

Obviously, the upstream face stresses are equal in absolute values, but with changed signs.

According to Equation 5.4, the deflection at the center of the plate is:

$$f = \alpha \cdot 0.353 \cdot 500^4 / (206000 \cdot 19^3) = 15.614\ \alpha$$

Table 5.4 gives, for $b/a = 2.5$, $\alpha = 0.0284$.

Hence, $f = 15.614 \cdot 0.0284 = 0.44$ mm.

5.3.3 EFFECTIVE WIDTH

The skin plate acts as a flange of stiffeners and beams. According to the NBR-8883 standard, the determination of the plate width to be used in the calculation of the composed section characteristics is made through Figures 5.4 and 5.5, in function of the bending moment distribution and the plate support lengths.

The effective plate width, on each side of the web, is equal to νB, where:

ν is the non-dimensional reduction factor, given in Figure 5-5 as a function of the L/B ratio, where L is the length of the moment zone of equal sign. L_I and ν_I are used in the zone between supports; and L_{II} and ν_{II}, in the support region;

B is half the span of the plate between two girders, or the overhang length.

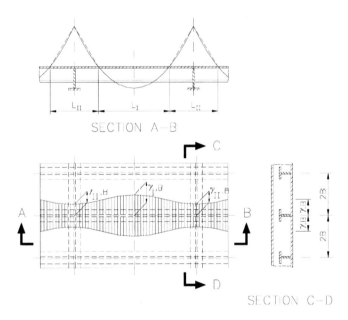

Fig. 5.4 Skin plate co-acting width (NBR-8883)

The NBR-8883 sets out an additional limitation of width of plates acting as curved beam flanges, such as the vertical beams of a segment gate.

That limit value is given by:

$$L_u = 1.56 \sqrt{R \ t} \qquad\qquad (5.5)$$

where:

R = plate curvature radius
t = plate thickness.

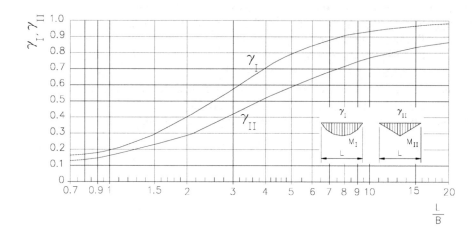

Fig. 5.5 Reduction factor (NBR-8883)

5.4 HORIZONTAL BEAMS

5.4.1 NUMBER OF BEAMS

When starting the gate leaf dimensioning, the designer faces the following problem: how many horizontal girders should the gate have? The determination of the number of beams follows a trial and error procedure. Initially, the number and dimensions of the beams are set out; then, stresses and deflections are calculated. If the results prove unsatisfactory, new values are determined and the stresses and deflections recalculated. This procedure is repeated until the desired result is attained.

To overcome this initial difficulty, the following empirical formula may be used (valid only for flat gates):

$$N = \frac{100h}{t} \sqrt{\frac{H_m}{2\sigma_{adm}}}$$ (5.6)

where:
  N = number of horizontal girders
  h = gate seal height, in meters
  t = skin plate thickness, in millimeters
  H_m = water head referred to the gate center, in meters
  σ_{adm} = allowable bending stress, in MPa.

The number of beams thus determined serves as a reference for the preliminary calculations of the gate leaf and only for that purpose. The final number of beams, as well as their spacing, is indicated only through the complete dimensioning of the structure.

The expression 5.6 derives from the equation of the plate bending stresses, wherein, to facilitate calculation, the following simplifications and assumptions were made:

a) the modules comprising the skin plate are rectangular, with the smallest support length a located in the vertical direction;

b) the b/a ratio of the rectangle support lengths is always above two, such that the k_{3x} coefficient in Table 5.3 be maximum and equal to 50, in the case of modules with rigid fixing of the four edges;

c) the pressure p has been replaced by the water head H_m in meters referred to the gate center;

d) the average beam spacing is assumed to be a, i.e., the smallest support length. So, the number of beams is given by $N = h/a$;

e) the plate bending stress has been replaced by the allowable bending stress, σ_{adm}.

Example 5.2 A 5.3 m high stoplog subjected to a water head of 40 m on the sill, is manufactured with ASTM A36 steel ($f_y = 248$ MPa). Determine the number of horizontal girders, knowing that the skin plate thickness is 16 mm.

Solution:

$h = 5.3$ m
$t = 16$ mm
$H_m = 40 - 5.3/2 = 37.35$ m

According to Table 5.1, the allowable bending stress is

$$\sigma_{adm} = 0.68 \cdot 248 = 168.64 \text{ MPa}$$

Therefore,

$$N = \frac{100 \cdot 5.3}{16} \sqrt{\frac{37.35}{2 \cdot 168.64}} = 11.02$$

Then, 11 horizontal girders will be needed. If the skin plate thickness is increased to 22 mm, the number of girders would be reduced to

$$N = \frac{100 \cdot 5.3}{22} \sqrt{\frac{37.35}{2 \cdot 168.64}} = 8.02$$

that is, eight girders.

Yet one must remember that a smaller number of girders will require stronger cross-sections because the beams are subjected to greater loads.

5.4.2 GIRDER DIMENSIONS

5.4.2.1 WEB THICKNESS

Once the number of horizontal girders is determined, the next step is the choice of their cross-section. For beams supported at the ends, the minimum thickness in the region of the support is given by the formula

$$t = \frac{F}{2\ h\ \tau_{adm}} \tag{5.7}$$

where:
   F = water load on the girder
   h = web depth
   τ_{adm} = allowable shear stress.

In the case of very wide gates, it is usually advantageous to build the web with plates thinner in the center region than at the ends of the beam, for shear is null in the center and maximum at the ends. The minimum recommended thickness for the web is 8 mm, though small gates may have webs with 6.5 mm thick plates.

5.4.2.2 WEB DEPTH

For structural reasons, the larger the head and the support span, the higher the horizontal girders. As a reference for the preliminary calculations of the characteristics of the cross-section of flat gate girders, see the empirical relations given in Table 5.5.

Table 5.5 Depth of Horizontal Girders

Head on Sill	Web Height
up to 15 m	from 1/12 up to 1/9 L
from 15 m to 3 0m	from 1/9 up to 1/7 L
over 30 m	from 1/7 up to 1/5 L

where L is the girder support length.

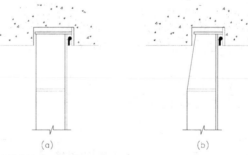

(a)   (b)

Fig. 5.6 Constant (a) and variable (b) girder depth

Girders may be designed with constant or variable depth along the span. Variable depth girders reduce the gate weight. Sloping of the horizontal girder depth at the ends permits reduction of the lateral slot dimensions.

5.4.2.3 FLANGES

Flanges are made from plates with thickness equal to or greater than that of the web. Their width can be taken equal to a fifth of the girder depth, in the preliminary calculations. When dimensioning the gate, the designer should consider as a whole the flange width, the spacing and depth of the horizontal girders, so as to facilitate access to the interior of the gate frame for welding. A minimum gap of 300 mm gap is recommended (see Figure 5.7) for gates with very high girders.

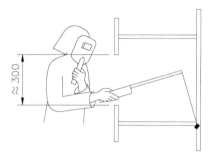

Fig. 5.7 Clearance for welding

5.4.3 ELASTIC STABILITY

5.4.3.1 COMPRESSION FLANGES

For built-up beams with loads applied in the plane of the web, enough safety should be built in against lateral buckling. Checking of lateral buckling may be done according to the DIN 4114 standard, which, in item 15.3, states that the compression flange is stable when the following condition is observed:

$$i_y > \frac{c}{40} \qquad\qquad (5.8)$$

where:
  i_y = major radius of gyration of the section formed by the flange and 1/5 of the web area;
  c = distance between the girder rigid points.

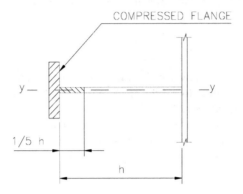

Fig. 5.8 Web participating area in lateral buckling resistance

Table 5.6 Buckling Coefficients for St 37 Steels (Yield strength f_y = 235.4 MPa) – DIN 4114

ω	ω +										ω
	0	1	2	3	4	5	6	7	8	9	
20	1.04	1.04	1.04	1.05	1.05	1.06	1.06	1.07	1.07	1.08	20
30	1.08	1.09	1.09	1.10	1.10	1.11	1.11	1.12	1.13	1.13	30
40	1.14	1 14	1 15	1.16	1.16	1.17	1.18	1.19	1.19	1.20	40
50	1.21	1.22	1.23	1.23	1.24	1.25	1.26	1.27	1.28	1.29	50
60	1.30	1.31	1.32	1.33	1.34	1.35	1.36	1.37	1.39	1.40	60
70	1.41	1.42	1.44	1.45	1.46	1.48	1.49	1.50	1.52	1.53	70
80	1.55	1.56	1.58	1.59	1.61	1.62	1.64	1.66	1.68	1.69	80
90	1.71	1.73	1.74	1.76	1.78	1.80	1.82	1.84	1.86	1.88	90
100	1.90	1.92	1.94	1.96	1.98	2.00	2.02	2.05	2.07	2.09	100
110	2.11	2.14	2.16	2.18	2.21	2.23	2.27	2.31	2.35	2.39	110
120	2.43	2.47	2.51	2.55	2.60	2.64	2.68	2.72	2.77	2.81	120
130	2.85	2.90	2.94	2.99	3.03	3.08	3.12	3.17	3.22	3.26	130
140	3.31	3.36	3.41	3.45	3.50	3.55	3.60	3.65	3.70	3.75	140
150	3.80	3.85	3.90	3.95	4.00	4.06	4.11	4.16	4.22	4.27	150
160	4.32	4.38	4.43	4.49	4.54	4.60	4.65	4.71	4.77	4.82	160
170	4.88	4.94	5.00	5.05	5.11	5.17	5.23	5.29	5.35	5.41	170
180	5.47	5.53	5.59	5.66	5.72	5.78	5.84	5.91	5.97	6.03	180
190	6.10	6.16	6.23	6.29	6.36	6.42	6.49	6.55	6.62	6.69	190
200	6.75	6.82	6.89	6.96	7.03	7.10	7.17	7.24	7.31	7.38	200
210	7.45	7.52	7.59	7.66	7.73	7.81	7.88	7.95	8.03	8.10	210
220	8.17	8.25	8.32	8.40	8.47	8.55	8.63	8.70	8.78	8.86	220
230	8.93	9.01	9.09	9.17	9.25	9.33	9.41	9.49	9.57	9.65	230
240	9.73	9.81	9.89	9.97	10.05	10.14	10.22	10.30	10.39	10.47	240
250	10.55										250

Table 5.7 Buckling Coefficients for St 52-3 Steels (Yield strength f_y = 353.2 MPa) – DIN 4114

ω	ω +										ω
	0	1	2	3	4	5	6	7	8	9	
20	1.06	1.06	1.07	1.07	1.08	1.08	1.09	1.09	1.10	1.11	20
30	1.11	1.12	1.12	1.13	1.14	1.15	1.15	1.16	1.17	1.18	30
40	1.19	1.19	1.20	1.21	1.22	1.23	1.24	1.25	1.26	1.27	40
50	1.28	1.30	1.31	1.32	1.33	1.35	1.36	1.37	1.39	1.40	50
60	1.41	1.43	1.44	1.46	1.48	1.49	1.51	1.53	1.54	1.56	60
70	1.58	1.60	1.62	1.64	1.66	1.68	1.70	1.72	1.74	1.77	70
80	1.79	1.81	1.83	1.86	1.88	1.91	1.93	1.95	1.98	2.01	80
90	2.05	2.10	2.14	2.19	2.24	2.29	2.33	2.38	2.43	2.48	90
100	2.53	2.58	2.64	2.69	2.74	2.79	2.85	2.90	2.95	3.01	100
110	3.06	3.12	3.18	3.23	3.29	3.35	3.41	3.47	3.53	3.59	110
120	3.65	3.71	3.77	3.83	3.89	3.96	4.02	4.09	4.15	4.22	120
130	4.28	4.35	4.41	4.48	4.55	4.62	4.69	4.75	4.82	4.89	130
140	4.96	5.04	5.11	5.18	5.25	5.33	5.40	5.47	5.55	5.62	140
150	5.70	5.78	5.85	5.93	6.01	6.09	6.16	6.24	6.32	6.40	150
160	6.48	6.57	6.65	6.73	6.81	6.90	6.98	7.06	7.15	7.23	160
170	7.32	7.41	7.49	7.58	7.67	7.76	7.85	7.94	8.03	8.12	170
180	8.21	8.30	8.39	8.48	8.58	8.67	8.76	8.86	8.95	9.05	180
190	9.14	9.24	9.34	9.44	9.53	9.63	9.73	9.83	9.93	10.03	190
200	10.13	10.23	10.34	10.44	10.54	10.65	10.75	10.85	10.96	11.06	200
210	11.77	11.28	11.38	11.49	11.60	11.71	11.82	11.93	12.04	12.15	210
220	12.26	12.37	12.48	12.60	12.71	12.82	12.94	13.05	13.17	13.28	220
230	13.40	13.52	13.63	13.75	13.87	13.99	14.11	14.23	14.35	14.47	230
240	14.59	14.71	14.83	14.96	15.08	15.20	15.33	15.45	15.58	15.71	240
250	15.83										250

If the former condition is not met, the flange is still considered stable, provided the maximum compression stress on the external fiber of the girder does not exceed the value

$$\frac{1.4\sigma_{adm}}{\omega} \qquad (5.9)$$

where:

σ_{adm} = allowable compression stress

ω = buckling coefficient given in Tables 5.6 and 5.7 as a function of the steel specification and the slenderness ratio.

The slenderness ratio is calculated by

$$\lambda = c/i_y \qquad (5.10)$$

5.4.3.2 WEB STABILITY

When dimensioning girders subjected to bending, it is convenient to space the flanges as much as possible and work with low thickness webs. However, the greater the ratio between the web depth and the thickness, the greater the possibility of buckling. The

German standard DIN 4114 recommends that, for built-up members, the web safety should be checked for values

$$\frac{h_w}{t_w} \geq 45 \qquad (5.11)$$

where:
   h_w = web depth
   t_w = web thickness.

The increase of the buckling strength of the girder web is achieved through the use of stiffeners. Rolled sections *I*, *C*, *H* and *L*-shaped are designed and manufactured with such dimensions that the web buckling by shear is not determinant, dispensing therefore with stiffeners.

Critical buckling stresses for girder webs are determined independently for the various rectangles contained between the stiffeners and the flanges. The rectangles are considered freely supported on the four edges. It is admitted that the shearing stress be uniformly distributed.

The value and sign of the ratio characterize the linear variation of the normal stress

$$\psi = \frac{\sigma_2}{\sigma_1} \qquad (5.12)$$

where σ_1 is the greater compression stress in the rectangle.

The sequence of calculation is:
a) calculation of the Euler reference stress, σ_E;
b) calculation of the ideal buckling stresses, σ_{fi} and τ_{fi};
c) calculation of the critical comparison stress, σ_{fic};
e) calculation of the buckling safety coefficient, v_f;
d) determination of the reduced comparison stress, σ_{rc}.

- Euler reference stress - is calculated by

$$\sigma_E = \frac{\pi^2 \; E \; t^2}{12(1-\mu^2)b^2} \qquad (5.13a)$$

where:
   t = web thickness
   b = rectangle height
   E = modulus of elasticity of steel
   μ = Poisson ratio.

For $E = 206000$ MPa and $\mu = 0.3$,

$$\sigma_E = 1.862 \cdot 10^5 \left(\frac{t}{b}\right)^2 \tag{5.13b}$$

- Ideal buckling stresses - are determined by the formulas

$$\sigma_{fi} = k_1 \sigma_E \qquad \text{(compression stress)} \tag{5.14}$$

and

$$\tau_{fi} = k_2 \sigma_E \qquad \text{(shear stress)} \tag{5.15}$$

where k_1 and k_2 are the buckling coefficients given in Table 5.8 as a function of the loading condition and the relation between the lengths of the rectangles,

$$\alpha = a/b \tag{5.16}$$

Table 5.8 Buckling Coefficients k_1 and k_2 (DIN 4114)

	1	2	3	4
1	LOADING	BUCKLING STRESS	VALIDITY RANGE	BUCKLING COEFFICIENT
2	COMPRESSIVE STRESSES LINEARLY DISTRIBUTED $0 \leqslant \psi \leqslant 1$	$\sigma_{fi} = k_1 \sigma_E$	$\alpha \geqslant 1$	$k_1 = \dfrac{8.4}{\psi + 1.1}$
			$\alpha < 1$	$k_1 = \left(\alpha + \dfrac{1}{\alpha}\right)^2 \dfrac{2.1}{\psi + 1.1}$
3	TENSION AND COMPRESSIVE STRESSES LINEARLY DISTRIBUTED, WITH PREDOMINANCE OF THE COMPRESSIVE STRESSES $-1 < \psi < 0$	$\sigma_{fi} = k_1 \sigma_E$		$k_1 = (1+\psi)k' - \psi k'' + 10\psi(1+\psi)$ WHERE k' IS THE BUCKLING COEFFICIENT FOR $\psi = 0$ (LINE 2) AND k'' IS THE BUCKLING COEFFICIENT FOR $\psi = -1$ (LINE 4)
4	TENSION AND COMPRESSIVE STRESSES LINEARLY DISTRIBUTED, WITH EQUAL END VALUES $\psi = -1$ OR PREDOMINANCE OF TENSION STRESSES $\psi < -1$	$\sigma_{fi} = k_1 \sigma_E$	$\alpha \geqslant \dfrac{2}{3}$	$k_1 = 23.9$
			$\alpha < \dfrac{2}{3}$	$k_1 = 15.87 + \dfrac{1.87}{\alpha^2} + 8.6\,\alpha^2$
5	SHEAR STRESSES EVENLY DISTRIBUTED	$\tau_{fi} = k_2 \sigma_E$	$\alpha \geqslant 1$	$k_2 = 5.34 + \dfrac{4.00}{\alpha^2}$
			$\alpha < 1$	$k_2 = 4.00 + \dfrac{5.34}{\alpha^2}$

- Critical comparison stress - is determined by

$$\sigma_{fic} = \frac{\sqrt{\sigma_1^2 + 3\tau^2}}{\frac{1+\psi}{4}\frac{\sigma_1}{\sigma_{fi}} + \sqrt{\left(\frac{3-\psi}{4}\frac{\sigma_1}{\sigma_{fi}}\right)^2 + \left(\frac{\tau}{\tau_{fi}}\right)^2}}$$

(5.17)

For σ_1 and τ, the greatest values of the compression and shear stresses occurring inside the rectangle are taken.

- Reduced comparison stress

The inelastic buckling occurs when the critical comparison stress proves greater than the yield strength of steel; σ_{fic} is then replaced by a reduced comparison stress (or actual critical stress), σ_{rc}, whose values are given in Table 5.9 for steels with $f_y = 235.4$ MPa and $f_y = 353.2$ MPa.

When σ_{fic} is lower than the yield strength, there is no reduction in the comparison stress and one adopts

$$\sigma_{rc} = \sigma_{fic}$$

(5.18)

Table 5.9 Comparative Reduced Stresses, σ_{rc} (DIN 4114)

σ_{fic} [MPa]	σ_{rc} [MPa]	
	St 37 Steel	St 52 Steel
188.4	188.4	188.4
196.2	194.5	196.2
206.0	199.7	206.0
215.8	203.8	215.8
225.6	206.9	225.6
235.4	209.5	235.4
245.3	211.7	245.3
255.1	213.7	255.1
264.9	215.2	264.9
274.7	216.7	274.7
284.5	217.9	284.4
294.3	219.1	291.7
313.9	220.9	301.9
333.5	222.4	308.9
353.2	223.7	314.2
372.8	224.7	318.6
392.4	225.6	322.2
412.0	226.4	325.0
431.6	227.1	327.5
451.3	227.7	329.5
470.9	228.2	331.4
490.5	228.7	333.0
539.6	229.6	336.1
588.6	230.2	338.4
637.7	230.8	340.3
686.7	231.3	341.8
784.8	232.1	343.9
981.0	232.9	346.5
1962.0	234.4	350.6
∞	234.4	353.2

- Buckling safety coefficient

The plate buckling safety is given by

$$\nu_f = \frac{\sigma_{rc}}{\sqrt{\sigma_1^2 + 3\tau^2}} \qquad (5.19)$$

The safety coefficient depends on the load cases and its minimum values are taken from Table 5.10.

Table 5.10 Safety Factors for Plate Buckling, ν_f

Load Case	ν_f(minimum)
Normal	1.35
Occasional	1.25
Exceptional	1.25

Example 5.3 Check the stability of the gate web shown in Figure 5.9, in the region between sections AA' and BB', knowing that the web thickness is 12.5 mm. The bending and shear stresses are:
- Section AA':
$\sigma_t = 113$ MPa
$\sigma_c = -91$ MPa
$\tau_A = 34$ MPa
 - Section BB':
$\sigma_t = 136$ MPa
$\sigma_c. = -111$ MPa
$\tau_B = 16$ MPa.
The gate is made from ASTM A36 steel ($f_y = 248$ MPa).

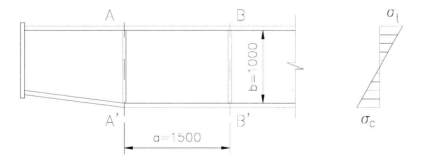

Fig. 5.9

Solution:

- Euler reference stress (Eq. 5.13a):

$$\sigma_E = \frac{206000\,\pi^2}{12\left(1\text{-}0.3^2\right)}\left(\frac{12.5}{1000}\right)^2 \qquad = 29.1 \text{ MPa (compression)}$$

The greatest compression stress occurs in section BB'. Therefore,

$\sigma_1 = \text{-}111$ MPa and $\sigma_2 = 136$ MPa

- Coefficient ψ (Eq. 5.12):

$\psi = 136/(\text{-}111) = \text{-}1.23$

The relation between the dimensions of rectangle ABB'A' is:

$\alpha = a/b = 1500/1000 = 1.5$

Table 5.8 gives, for $\psi = \text{-}1$ and $\alpha > 2/3$,

$k_1 = 23.9$

and

$$k_2 = 5.34 + \frac{4}{\alpha^2} = 5.34 + \frac{4}{1.5^2} = 7.12$$

According to Equations. 5.14 and 5.15, the ideal buckling stresses will be:

$\sigma_{fi} = 23.9 \cdot 29.1 = 695.5$ MPa (compression)
and
$\tau_{fi} = 7.12 \cdot 29.1 = 207.2$ MPa

- Critical comparison stress (Eq. 5.17):

$$\sigma_{fic} = \frac{\sqrt{\left(\text{-}111\right)^2 + 3\cdot 34^2}}{\dfrac{1+\left(\text{-}1.23\right)}{4}\,\dfrac{\left(\text{-}111\right)}{\left(\text{-}695.5\right)} + \sqrt{\left[\dfrac{3\text{-}\left(\text{-}1.23\right)}{4}\,\dfrac{\left(\text{-}111\right)}{\left(\text{-}695.5\right)}\right]^2 + \left(\dfrac{34}{207.2}\right)^2}} = 555.5 \text{ MPa}$$

As σ_{fic} is greater than the yield strength of steel, it should be replaced by the reduced comparison stress. In Table 5.9, by interpolation, we have for the steel in question,

$\sigma_{rc} = 229.8$ MPa.

Therefore, the buckling safety coefficient is (Eq. 5.19):

$$v_f = \frac{229.8}{\sqrt{(-111)^2 + 3 \cdot 34^2}} = 1.83 \rangle \ 1.35$$

As the safety coefficient is greater than 1.35 (see Table 5.10), the web is stable and does not require stiffeners.

5.4.3.3 STIFFENERS

It is possible to increase the critical buckling stress of a plate without increasing its thickness, through the use of stiffeners adequately provided. Generally, the installation of stiffeners obeys the following recommendations:
. pure compression: longitudinal stiffeners placed symmetrically in relation to the longitudinal axis;
. simple bending: longitudinal stiffeners placed at a fourth or a fifth of the web depth, near the compression flange, on one side of the web plate;
. shear: transverse stiffeners, dividing the web into square plates with each side equal to the web depth. The DIN 4114 standard presents a number of tables for determination of the minimum moment of inertia of the stiffeners.

5.4.4 SIMPLE BENDING OF BEAMS

The load diagram of horizontal girders of flat gates (fixed-wheel, slide, stoplog and caterpillar) is shown in Figure 5.10.

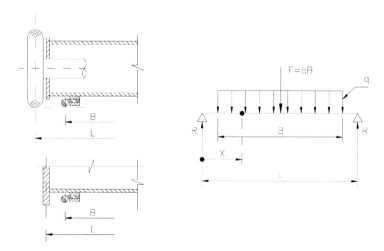

Fig. 5.10 Horizontal girder loading

The bending moment at any point, such that

$$L/2 \geq x > (L - B)/2$$

is given by

$$M_x = Rx - \frac{F}{2B}\left[x - \frac{1}{2}(L - B) \right]^2 \qquad (5.20)$$

and its maximum value occurs at the midspan ($x = L/2$), being determined by the equation

$$M = (2L-B)F/8 \qquad (5.21)$$

where:
  R = support reaction = F/2
  x = distance between the support and the considered point
  F = water load acting on the girder
  B = seal span
  L = support length
  q = load per unit length = F/B.

The horizontal girder deflection under load should be limited so as not to affect the safety, the movement and, in the case of upstream seals, the gate water tightness. Maximum deflection occurs at midspan and is determined by

$$f = \frac{5\ F\ L^3}{384\ E\ I} \qquad (5.22)$$

where:
  F = water load on the girder
  L = support length
  E = modulus of elasticity of steel
  I = moment of inertia of girder cross-section.

Maximum deflection of the main girders is usually limited to 1/750 of the support length. The vertical end girders of fixed-wheel gates are supported on the wheel pins and designed to resist the water loads transmitted by the horizontal girders. In case of more than two wheels per vertical end girder, the system is statically undetermined and the girder is designed as a continuous beam.

5.5 SEGMENT GATE

5.5.1 SKIN PLATE

The skin plate of segment gates is made up of curved plates joined by butt welds. The minimum recommended thickness is 8 mm, except for small weir gates, where 6.5 mm plates may be used. Plates with different thicknesses are used on high gates, with the thickest plate located on the lower leaf portion, where the water pressure is greater. Plates more than 18 mm thick are seldom used in weir gates. In the manufacture of the spillway gates of Itaipu, with a 20 m span and a height of 21.34 m, only plates with 14 mm and 12 mm thickness were used in the skin plate.

For ease of manufacture and transportation, the skin plate is usually subdivided into horizontal elements. The arc length of the various elements should be the greatest possible (but within the transportation limits) to reduce the amount of field joints. Butt welding of the skin plate and vertical girder carries out the union of the elements on site.

Thanks to the great facility of inspection and maintenance, in spillway gates there is no need to provide additional thickness of the skin plate for corrosion.

5.5.2 GATE FRAMING

5.5.2.1 GIRDER ARRANGEMENT

Strengthening of the skin plate is made with the aid of horizontal and vertical girders and stiffeners. According to the arrangement of the girders, two classical constructions are known (see Figure 5.11).

In Figure 5.11a, the gate framing consists substantially of horizontal girders supported on two vertical main beams. The auxiliary horizontal beams are arranged with variable spacing so that all are subjected to the same part of the water thrust, which greatly facilitates their dimensioning and manufacture.

For the design shown in Figure 5.11b, the gate framing comprises a series of auxiliary vertical beams, made of curved rolled sections, supported on the main horizontal girders. In both designs, vertical and horizontal stiffeners further strengthen the skin plate.

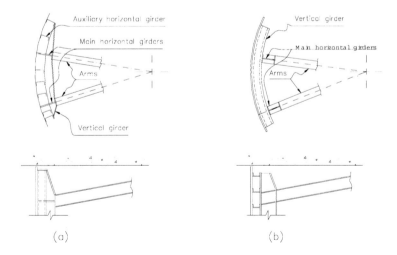

Fig. 5.11 Typical arrangements of girders

On modern gates, the framing construction based on the system of horizontal girders (Fig. 5.11a) is predominant. On low weir gates, or on those whose height is less than the span, it may be convenient to adopt the system of vertical beams. The advantage of this design, however, should be demonstrated through an economic comparative study between the two solutions.

5.5.2.2 HORIZONTAL BEAMS

Dimensioning of horizontal beams with both ends overhanging usually falls into one of the following cases:

a) rounded connections - occurs when the length of the overhanging section is made equal to 0.225 times the beam length (see Figure 5.12). Under these conditions, the elastic tangent at the supports will be zero, meaning that there is no transmission of the moment to the supports. The bending moment is maximum at the supports and equal to about twice that of the midspan, in absolute value.

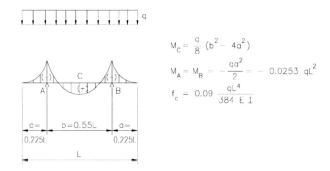

$$M_C = \frac{q}{8}(b^2 - 4a^2)$$

$$M_A = M_B = -\frac{qa^2}{2} = -0.0253\, qL^2$$

$$f_c = 0.09\, \frac{qL^4}{384\, E\, I}$$

Fig. 5.12 Horizontal girder with rounded connections

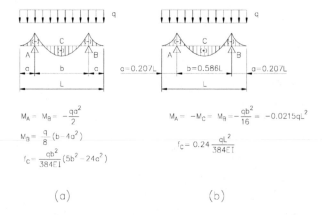

$$M_A = M_B = -\frac{qa^2}{2}$$

$$M_B = \frac{q}{8}(b - 4a^2)$$

$$f_c = \frac{qb^2}{384EI}(5b^2 - 24a^2)$$

$$M_A = -M_C = M_B = -\frac{qb^2}{16} = -0.0215qL^2$$

$$f_c = 0.24\, \frac{qL^2}{384EI}$$

(a)   (b)

Fig. 5.13 Horizontal girder with rigid connections

b) rigid connections - occurs whenever the length of the overhanging differs from 0.225 times the beam length. In the particular case of $a = 0.207L$ (see Figure 5.13b), the bending moment at midspan will be equal in magnitude (but with opposite sign) to that of the support. Deflection at midspan is well above that corresponding to the above case of rounded connections.

The behavior of the main horizontal girders connected to the radial arms by rigid connections does not follow exactly what has been described in b, and their dimensioning is made along that of the arms, through analysis of the rigid frame so constituted (see item 5.5.3.3).

5.5.2.3 VERTICAL BEAMS

Support of vertical beams is made on the main horizontal girders, where the radial arms are also connected. In more recent designs, the use of two pairs of radial arms per gate prevails, which results in vertical beams with two supports.
In the choice of the arms geometry in the vertical plane (which also defines the position of the main horizontal girders), two methods may be used:

a) arms equally loaded in the axial direction - this situation occurs when the direction of the maximum water thrust coincides with the bisectrix of the angle between the upper and lower arms, in the case of gates with a two pairs of radial arms;
b) equal bending moments on the main vertical girder supports.

Fig. 5.14 Spillway segment gate, Itaipu Dam

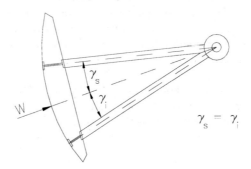

Fig. 5.15 Equally loaded arms

On submerged gates, the arms position determined according to Tables 5.11 and 5.12 results in equal bending moments on the supports.

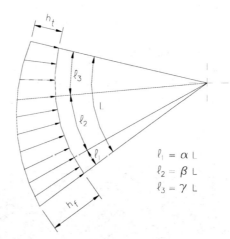

Fig. 5.16 Submerged segment gate with two pair of arms
(L is the skin plate "wet" length)

Table 5.11 Coefficients for Positioning of Arms (see Figure 5.16)

h_t / h_f	α	β	γ
0.1	0.1536	0.5114	0.3350
0.2	0.1634	0.5348	0.3018
0.3	0.1718	0.5496	0.2786
0.4	0.1790	0.5596	0.2614
0.5	0.1858	0.5656	0.2486
0.6	0.1918	0.5700	0.2382
0.7	0.1974	0.5728	0.2298
0.8	0.2028	0.5740	0.2232
0.9	0.2078	0.5748	0.2174

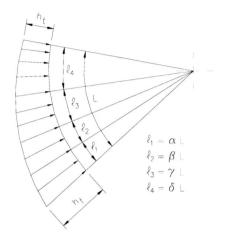

Fig. 5.17 Submerged segment gate with three pair of arms
(L is the skin plate "wet" length)

Table 5.12 Coefficients for Positioning of Arms (see Figure 5.17)

h_t / h_f	α	β	γ	δ
0.1	0.1046	0.3024	0.3462	0.2468
0.2	0.1111	0.3165	0.3567	0.2157
0.3	0.1167	0.3262	0.3613	0.1958
0.4	0.1216	0.3338	0.3628	0.1818
0.5	0.1262	0.3394	0.3627	0.1717
0.6	0.1304	0.3438	0.3619	0.1639
0.7	0.1343	0.3475	0.3605	0.1577
0.8	0.1381	0.3505	0.3587	0.1527
0.9	0.1416	0.3531	0.3568	0.1485

In the particular case of weir gates, the arms position is defined by:

. gates with two pairs of arms:

$$l_1 = 0.1414 \, L \tag{5.23a}$$
$$l_2 = 0.4734 \, L \tag{5.23b}$$
$$l_3 = 0.3852 \, L \tag{5.23c}$$

. gates with three pairs of arms:

$$l_1 = 0.0964 \, L \tag{5.24a}$$
$$l_2 = 0.2804 \, L \tag{5.24b}$$
$$l_3 = 0.3233 \, L \tag{5.24c}$$
$$l_4 = 0.2999 \, L \tag{5.24d}$$

The vertical beams may be dimensioned as straight ones, supported on the main horizontal girders and subjected to a trapezoid load (submerged gates) or triangular load (weir gates).

For weir gates with two pairs of radial arms, the load diagram is shown in Figure 5.18.

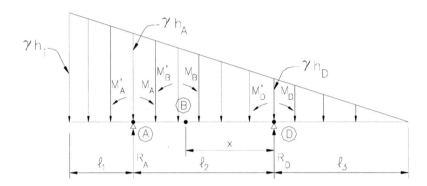

Fig. 5.18 Loading on vertical beams of spillway segment gate

The bending moments at the supports, per unit width, are:

$$M_A' = \gamma \frac{l_1^2}{6}(2h_i + h_A) \tag{5.25}$$

$$M_D' = \gamma \frac{l_3^2}{6}h_D \tag{5.26}$$

where:
  h_i = depth corresponding to lower seal
  h_A = depth corresponding to support A
  h_D = depth corresponding to support D
  γ = specific weight of water = 9.81 kN/m^3

The bending moment on any point B between the supports and at a distance x from support D is:

$$M_B' = \gamma \left[\frac{x\, h_D}{2}(x - l_2) + \frac{(h_A - h_D)\, x}{6\, l_2}(x^2 - l_2^2) \right] + \frac{M_A' + M_D'}{2} \tag{5.27}$$

An investigation is made on the above equation to determine the point corresponding to the maximum bending moment in the length between supports.

For submerged gates, the loading is trapezoidal and the calculation method is similar to that of the weir gate. The only difference is on account of the bending moment on support

D, since the overhanging length to the right of D is subjected to a trapezoidal, not a triangular loading. In this situation, the bending moment at D is determined by

$$M_D' = \gamma \frac{l_3^2}{6}(2h_t + h_D)$$

(5.28)

where
  h_t = depth corresponding to the upper seal.

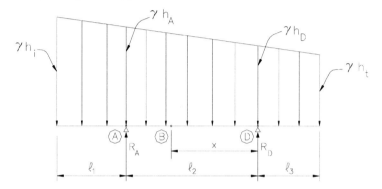

Fig. 5.19 Loading on vertical beams of submerged segment gate

The bending moments M'_A and M'_B are determined as indicated for weir gates. For gates with three or more pairs of arms, the vertical girders are calculated as continuous beams, using the equation of the three moments.

Example 5.4 The segment gate of Figure 5.20 has a span of 12 m, height of 16 m and radius of 15 m, and is subjected to a water head of 15 m on the sill. The gate has two pairs of radial arms connected to two main vertical girders. Determine geometry of the arms so as to have equal bending moments at the vertical girder supports. Also, calculate the maximum bending moment in the vertical girder and the point where it occurs.

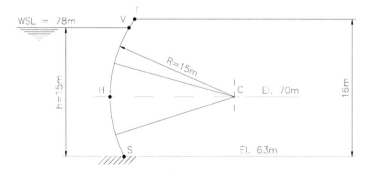

Fig. 5.20

Solution:

- Developed length of the skin plate (submerged length only):

The length of the arc HV lying above the hinge elevation is:

$$L_s = \beta_s R = (\text{arc sin VT/R}) R$$
$$VT = El. 78 - El. 70 = 8 \text{ m}$$
$$R = 15 \text{ m}$$
$$\beta_s = \text{arc sin VT/R.} = \text{arc sin } 8/15 = 32.231 \text{ degrees} = 0.563 \text{ rad}$$
$$\therefore L_s = 0.563 \cdot 15 = 8.438 \text{ m}$$

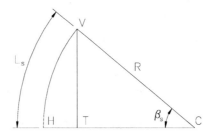

Fig. 5.21

The length of the arc HS will be:
$$L_i = \beta_i R = (\text{arc sin QS/R}) R$$
$$QS = El. 70 - El. 63 = 7 \text{ m}$$
$$\beta_i = \text{arc sin QS/R.} = \text{arc sin } 7/15 = 27.818 \text{ degrees} = 0.486 \text{ rad}$$
$$\therefore L_i = 0.486 \cdot 15 = 7.283 \text{ m}$$

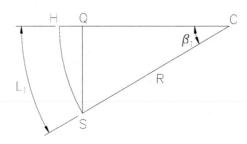

Fig. 5.22

Therefore, the developed length of the skin plate is
   $L = L_s + L_i = 8.438 + 7.283 = 15.721$ m.

Equations 5.23a, 5.23b and 5.23c define the arms position to have equal bending moments in the vertical girder supports:
   $l_1 = 0.1414 \, L = 0.1414 \cdot 15.721 = 2.223$ m
   $l_2 = 0.4734 \, L = 0.4734 \cdot 15.721 = 7.442$ m
   $l_3 = 0.3852 \, L = 0.3852 \cdot 15.721 = 6.056$ m

Thus,

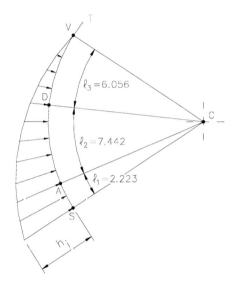

Fig. 5.23

- Bending moments in the vertical girders

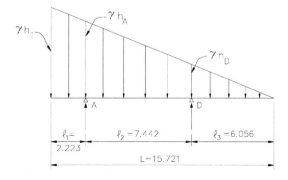

Fig. 5.24

Dimensioning the vertical beams as straight ones, gives:

$$\frac{\gamma \; h_i}{\gamma \; h_A} = \frac{L}{l_2 + l_3}$$

$\therefore h_A = 15 \cdot 13.498/15.721 = 12.879$ m

$$\frac{\gamma \; h_i}{\gamma \; h_D} = \frac{L}{l_3}$$

$\therefore h_D = 15 \cdot 6.056/15.721 = 5.778$ m

Considering, initially, the gate with unit width, then,

- Bending moment at support A (Eq. 5.25)

$M_A' = 9.81 \cdot 2.223^2 (2 \cdot 15 + 12.879)/6 = 346.45$ kN-m/m

- Bending moment at support D (Eq. 5.26)

$M_D' = 9.81 \cdot 5.778 \cdot 6.056^2/6 = 346.47$ kN-m/m

$\therefore M_A' \approx M_D'$.

Equation 5.27 determines the bending moment at any point B situated between the supports:

$$M_B' = 9.81 + \left[\frac{5.778 \; x}{2}(x\text{-}7.442) + \frac{(12.879\text{-}5.778) \; x}{6 \cdot 7.442}\left(x^2\text{-}7.442^2\right) \right] + \frac{346.45 + 346.47}{2} =$$
$$= 28.34x^2 - 210.91x + 1.56x^3 - 86.4x + 346.46 =$$
$$= 1.56x^3 + 28.34x^2 - 297.31x + 346.46$$

Search of the point where the maximum bending moment M_B':

Deriving the above equation with respect to x and making it equal to zero, results:

$4.68x^2 + 56.68x - 297.31 = 0$

The roots of the equation are $x_1 = 3.95$ and $x_2 = -16.07$.

Hence, the root that satisfies the above equation is $x_1 = 3.95$, and the maximum bending moment is obtained by replacing x by 3.95 in the equation of M_B':

$M_B' = 1.56 \cdot 3.95^3 + 28.34 \cdot 3.95^2 - 297.31 \cdot 3.95 + 346.46 = -289.6$ kN-m/m.

The following bending moment diagram can then be set out:

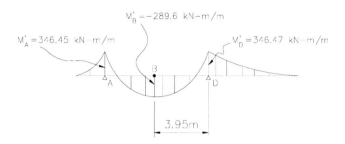

Fig. 5.25

Since the bending moments have been calculated for a unit width gate, its is necessary to multiply them by the half the span, to give the moment in each girder:

$M_A = M_A$' $B/2 = 346.45 \cdot 12/2 = 2078.7$ kN-m
$M_B = M_B$' $B/2 = (-289.6)12/2 = -1737.6$ kN-m
$M_D = M_D$' $B/2 = 346.47 \cdot 12/6 = 2078.8$ kN-m.

5.5.3 RADIAL ARMS

5.5.3.1 AXIAL LOADS ON THE ARMS

The radial arms of segment gates transfer the water thrust acting on the gate leaf to the bearings. The distribution of the water thrust to the arms in gates with two pairs of arms is made as follows:

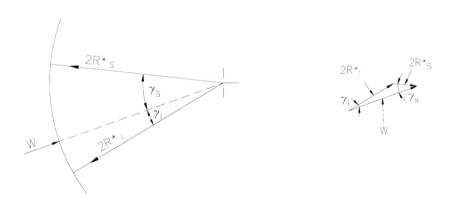

Fig. 5.26 Segment gate arm reactions

a) Case of radial arms parallel to pier face:

- axial load on each upper arm

$$R_s* = \frac{W}{2} \frac{\sin \gamma_i}{\sin(\gamma_i + \gamma_s)} \tag{5.29}$$

- axial load on each lower arm

$$R_i* = \frac{W}{2} \frac{\sin \gamma_s}{\sin(\gamma_i + \gamma_s)} \tag{5.30}$$

b) Case of sloped radial arms:

- axial load on each upper arm

$$R_s = \frac{R_s*}{\cos \omega} = \frac{W}{2 \cos \omega} \frac{\sin \gamma_i}{\sin(\gamma_i + \gamma_s)} \tag{5.31}$$

- axial load on each lower arm

$$R_i = \frac{R_i*}{\cos \omega} = \frac{W}{2 \cos \omega} \frac{\sin \gamma_s}{\sin(\gamma_i + \gamma_s)} \tag{5.32}$$

where ω is the angle between the arm and the pier face.

5.5.3.2 BEARING LOADS

For gates with sloped radial arms, the load transmitted to the bearing may be divided into two orthogonal components, F_n and F_r whose values are:

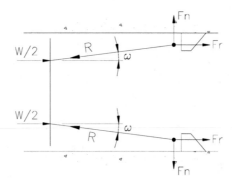

Fig. 5.27 Segment gate bearing loads

$$F_n = R \sin w \qquad\qquad (5.33)$$

$$F_r = R \cos w \qquad\qquad (5.34)$$

On the other hand,

$$R = \frac{W}{2 \cos \omega} \qquad\qquad (5.35)$$

Hence,

$$F_n = \frac{W}{2 \cos \omega} \sin \omega = \frac{W}{2} \tan \omega \qquad\qquad (5.36)$$

$$F_r = \frac{W}{2 \cos \omega} \cos \omega = \frac{W}{2} \qquad\qquad (5.37)$$

Example 5.5 Determine the axial loads on the arms and the axial (F_n) and radial (F_r) components of the thrust on the bearings of the segment gate of Figure 5.28, knowing that the maximum water thrust on the gate is $W = 18$ MN and its direction makes an angle of 10 degrees with the upper arms and 18 degrees with the lower arms. The gate arms make an angle of 8 degrees with the pier face.

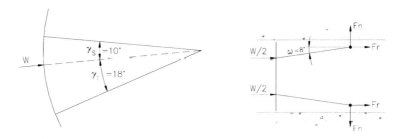

Fig. 5.28

Solution:

- Axial loads on each arm:

Upper arms (Eq. 5.31):

$$R_s = \frac{18000}{2 \cos 8^\circ} \; \frac{\sin 18^\circ}{\sin\left(18^\circ + 10^\circ\right)} = 5982 \text{ kN}$$

Lower arms (Eq. 5.32):

$$R_i = \frac{18000}{2 \cos 8^\circ} \; \frac{\sin 10^\circ}{\sin\left(18^\circ + 10^\circ\right)} = 3362 \text{ kN}$$

- Bearing loads:

The resultant load on each bearing has the following components:

$$F_n = \tfrac{1}{2}\ 18000 \tan 8° = 1265 \text{ kN}$$

and

$$F_r = \tfrac{1}{2}\ 18000 = 9000 \text{ kN}.$$

5.5.3.3 BUCKLING CHECK

The arms of segment gates should be checked against buckling caused by the combination of axial and bending loads, both on the vertical plane and on the planes formed by the main horizontal girders and their respective arms.
   The following loads should be considered:
a) axial compression due to hydraulic loads;
b) own weight of arms;
c) friction on the bearings;
d) operating forces due to action of hoists;
e) bending moments on supports (rigid fixing) and bearings.

The bending moment caused by friction on the bearings acts on the vertical plane and is distributed to the arms proportionally to their rigidity. The resisting moment on segment gate bearings is calculated as shown in item 9.3.
   During movement of the gate, the operating devices introduce loads on the arms, chiefly in the axial direction, which must be taken into account in their dimensioning. Figure 5.29 illustrates that influence.

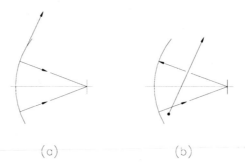

(a)   (b)

Fig. 5.29 Arm loads caused by the operating mechanism
(a) Cables or chains; (b) Hydraulic hoists, cables or chains

The arms bending moments acting on the planes formed by these and the main horizontal girders are determined as a function of the bearing type and the connection between the arms and the girder. To determine the bending moments, the rigid frame formed by the neutral axis of the horizontal girder and the axis of the arms should be analyzed. Figure 5.30 shows the main types of rigid frames, their loading and the bending moments.

In cases a and c, the hinge axes are provided with anti-friction bearings or cylindrical bushings, resulting in a rigid fixing in the rigid frame plane. Cases b and d, apply to gates with pivoted bearing (spherical plain bearing, e.g.). Loads on bearings are calculated as shown in item 5.5.3.2.

In the design shown in Figure 5.30c, with $a = 0.225\ L$, the bending moment M_{BI} is equal to zero, as well as M_D. Thus, the arms are subjected only to axial loads.

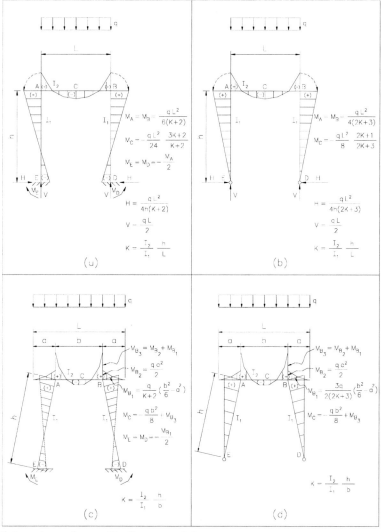

Fig. 5.30 Girder and arms frame – Bending moment diagrams
(a) parallel arms and cylindrical bushings; (b) parallel arms and spherical bushings; (c) sloped arms and cylindrical bushings; (d) sloped arms and spherical bushings

On the vertical plane, the arms are braced in order to reduce the buckling length. The arms bracing on the plane formed by the arms and the horizontal girders is a complex task, mainly on very wide gates. So, the arm sections are usually arranged with the major inertia in the horizontal direction corresponding to the arm's greatest unsupported length.

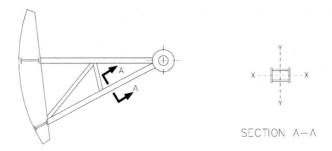

SECTION A—A

Fig. 5.31 Vertical bracing of gate arms

Stability of arms may be checked according to the DIN 4114 standard, which establishes the following relation:

$$\omega \, \frac{F}{A} \, + \, 0.9 \, \frac{M}{W_c} \leq \sigma_{adm} \tag{5.38}$$

where:

   ω = buckling coefficient given by Tables 5.6 and 5.7 as a function of the steel characteristics and the slenderness ratio λ
   F = compression load
   A = rough cross-section area
   M = bending moment
   W_c = section modulus (compression)
   σ_{adm} = allowable compression stress.

The arm slenderness ratio is given by:

$$\lambda = l_f / i \tag{5.39}$$

where:
   l_f = buckling length
   i = radius of gyration = $\sqrt{I / A}$ $\tag{5.40}$

The buckling length is measured on the arm axis. On the rigid frame plane formed by the arms and the horizontal girder, l_f is taken equal to the distance between the hinge axis and the neutral axis of the horizontal girder. On the vertical plane, the buckling length is equal to the distance between the bracing connections, if any.

Chapter 6

Embedded Parts, Guides and Supports

6.1 SLOTS AND NICHES

Slide, stoplog, fixed-wheel, caterpillar, roller and *Stoney* gates bear on embedded parts installed in slots built into the sidewalls. Slots usually house the following gate support and guide elements: wheel and sliding tracks; seal seats; side guides; counterguides; concrete edge protectors; and, eventually, slot lining.

Direct installation of embedded parts in the first stage concrete is usually avoided because it hinders the attainment of the required tolerances for the proper positioning of the elements. Thus, most times, the embedded parts are installed within niches or pits built in the first stage concrete and, after their adjustment and placement of forms, enveloped by the second stage concrete.

Positioning of the embedded parts components within the niches is made either by anchor bolts or adjusting screws welded to re-bars or anchor plates embedded in the first stage concrete. Setting of the embedded parts is made, whenever possible, in the two main directions, that is, in the flow direction and orthogonal to it.

The second stage concreting is made with the aid of rigid metal or wood forms. If the slot is completely lined, the lining may act as a form for concreting, provided it has been duly designed for that purpose. It should be emphasized that the slot lining should be manufactured in sections, in order to permit access to all elements of the embedded parts being erected. The final installation of the sections is carried out only after conclusion of the position adjustment of the embedded parts. In the design of the embedded parts, holes should be provided in the stiffeners and horizontal ribs, thereby making it easier for pouring and vibration of the concrete.

Gate slots disrupt the boundary lines, create vortices and provoke the separation of the flow at the slot boundaries, causing low pressures in the vicinity of the slot. Figure 6.1 shows the flow past a rectangular slot.

Flow separation at the downstream slot edge creates a low-pressure zone in A. The flow stagnation in B and vortices arising in C may be observed [1]. The combination of high flow speed with severe low pressures in the downstream wall of the slot may cause erosion and cavitation.

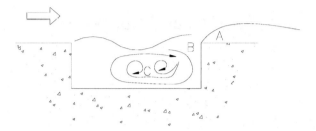

Fig. 6.1 Flow past a rectangular slot

Offsetting the downstream slot edge (see Figure 6.2) substantially improves the local conditions of the flow, eliminating or reducing the low pressure points near the wall.

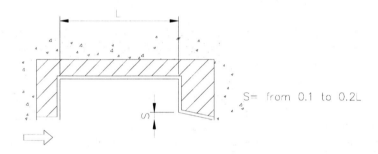

Fig. 6.2 Offset downstream corner

Sharma [2] recommends setting an offset to the downstream end of the slot equivalent to $10 \approx 20$ per cent of the slot length. Since the flow speed in the vicinity of the slot of high-head gates is very high, its downstream edge should be rounded and its junction with the sidewall made at a slope of 1 to 10 or flatter. In low-head gates (weir gates, for example), there is no need to offset the downstream wall.

The slot final dimensions should be kept as small as possible, so as to minimize hydraulic flow disturbance and to avoid trapping of debris. Niches constructed in the first stage concrete, in turn, should be wide enough to allow welding of the anchor bolts to the anchor plates, as well as providing adequate access to the adjusting nuts.

A survey carried out by the author in about 44 gates with spans from 5 m to 20 m, indicates the following relations between the main dimensions of the niches built in the first stage concrete and the gate span:

Table 6.1 Ratio between Niche Dimensions and Gate Span (see Figure 6.3)

Gate type	Length/Span (M/B)			Depth/Span (N/B)		
	Minimum	Mean	Maximum	Minimum	Mean	Maximum
Spillway stoplog	0.10	0.12	0.15	0.04	0.05	0.07
Submerged stoplog	0.15	0.23	0.27	0.05	0.09	0.14
Fixed-wheel gate	0.15	0.27	0.40	0.06	0.12	0.22

Fig. 6.3 Basic dimensions of lateral slots

The final slot profile is usually determined by the gate supplier as a function of the design of the embedded parts. The slot length depends on the support type of the gate (through wheels - fixed-wheel gates; or end girders - stoplogs, for example) and the web height at the ends of the horizontal girders. Fixed-wheel gates with end-supported wheel pins require deeper slots than those provided with cantilevered pins. In some cases, the slot occupies all the width of the niche built in the first stage concrete (see Figures 2.77 and 2.86).

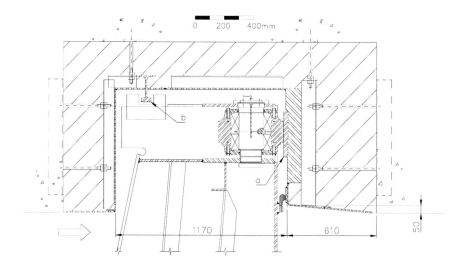

Fig. 6.4 Slot section of the intake fixed-wheel gate, Itumbiara Power Plant (KRUPP)
(a) wheel track; (b) side guide

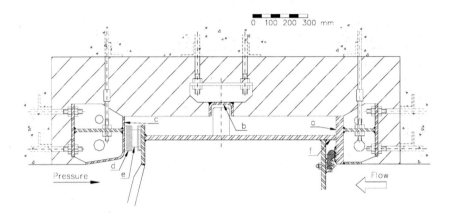

Fig. 6.5 Slot section of the draft tube stoplogs, São Simão Power Plant (DEDINI)
(a) slide track; (b) side guide; (c) counterguide; (d) wedge; (e) spring; (f) vertical end girder

6.2 WHEEL TRACK

6.2.1 BEAM ON AN ELASTIC FOUNDATION

The wheel track can be dimensioned as an infinitely long beam resting on an elastic continuum and subjected to concentrated loads transmitted by the wheels. The reaction of the foundation upon the beam is proportional to the beam deflection. So, the reaction of the foundation per unit length is given by:

$$q = - k\, y \qquad\qquad (6.1)$$

where y is beam deflection and k the modulus of the foundation. Biot [3] developed two methods for determining the modulus of the foundation. In the first, the beam is resting on top of a pillar infinitely high and with a width equal to that of the beam base, characterizing a two-dimensional foundation. Thus, we have:

$$k = 0.28\, E_c \sqrt[3]{\frac{E_c\, b^4}{E_a\, I}} \qquad\qquad (6.2a)$$

where
E_c = modulus of elasticity of the foundation
E_a = modulus of elasticity of the beam
b = support width
I = moment of inertia of the cross section of the beam.

In the second case, Biot analyses the behavior of a beam supposed to rest on a foundation wider than the beam base. The foundation is considered a three-dimensional elastic half space, and the modulus of the foundation is given by:

$$k = 0.91 \left[\frac{1}{C\,(1\text{-}v^2)} \frac{E_c\ b^4}{E_a\ I} \right]^{0.11} \frac{E_c}{C\,(1\text{-}v^2)} \tag{6.2b}$$

where:
  C is a coefficient varying from 1 for uniform pressure distribution to 1.13 for uniform deflection; and
  v is the Poisson ratio of the foundation, taken equal to 0.20 for concrete.

Analyzing the usual construction of wheel tracks, one notices that the foundation width, on one side, exceeds slightly that of the beam base and, on the other side, is much greater. Therefore, the use of any of the above formulas will result in approximate values of the modulus of the foundation. The Brazilian standard NBR-8883 recommends this formula:

$$k = 0.4\ E_c\ \sqrt[3]{\frac{E_c\ b^4}{E_a\ I}} \tag{6.2c}$$

This equation was determined by making the expression, which defines the concrete bearing pressure, according to the Andree-Fricke theory, equal to the corresponding expression of the beam on an elastic foundation.

According to Hetényi [4], the general equations of the theory of the beam resting on an elastic foundation are:

- deflection curve:

$$y = \frac{P\ \beta}{2\ k} e^{-\beta x} \left(\cos\ \beta x + \sin\ \beta x \right) \tag{6.3}$$

- angular deflection:

$$\theta = -\frac{P\beta^2}{k} e^{-\beta x}\ \sin \beta x \tag{6.4}$$

- bending moment:

$$M = \frac{P}{4\beta} e^{-\beta x} \left(\cos \beta x - \sin \beta x \right) \tag{6.5}$$

- shearing force:

$$Q = -\frac{P}{2} e^{-\beta x} \cos \beta x \tag{6.6}$$

where:
  P = concentrated load
  x = distance between the load P and the point considered
  β = coefficient of the foundation, given by

$$\beta = \sqrt[4]{\frac{k}{4E_a \; I}}$$

(6.7)

Note: In the calculation of sin βx and cos βx, βx is given in radians.
Introducing in the above equations the notations

$A_{\beta x} = e^{-\beta x} (\cos \beta x + \sin \beta x)$ (6.8)

$B_{\beta x} = e^{-\beta x} \sin \beta x$ (6.9)

$C_{\beta x} = e^{-\beta x} (\cos \beta x - \sin \beta x)$ (6.10)

$D_{\beta x} = e^{-\beta x} \cos \beta x$ (6.11)

One may rewrite them as

$$y = \frac{P \; \beta}{2 \; k} A_{\beta x}$$

(6.3a)

$$\theta = - \frac{P\beta^2}{k} B_{\beta x}$$

(6.4a)

$$M = \frac{P}{4 \; \beta} C_{\beta x}$$

(6.5a)

$$Q = - \frac{P}{2} D_{\beta x}$$

(6.6a)

The curves representing these equations are shown in the Figure 6.6.

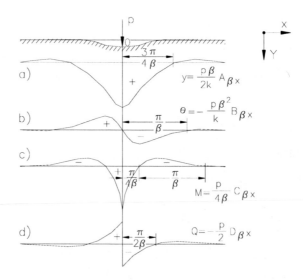

Fig. 6.6

The overlapping effect due to the wheel loads should be considered in the wheel track dimensioning (see Examples 6.1 and 6.2).

The bending stress of the wheel track is:

$$\sigma_f = \frac{M}{W} = \pm \frac{P}{4W\beta} \, C_{\beta x} \qquad (6.12)$$

The shear stress is calculated by the formula

$$\tau = \frac{Q}{A_a} = - \frac{P}{2A_a} D_{\beta x} \qquad (6.13)$$

where A_a is the area of the wheel track subjected to shearing.

Assuming a uniform pressure distribution under the beam base, the concrete bearing pressure will be given by

$$p_o = \frac{q}{b} = - \frac{ky}{b} = - \frac{kP\beta}{2bk} A_{\beta x} = - \frac{P\beta}{2b} A_{\beta x} \qquad (6.14)$$

The bearing pressure shall not exceed the comparative stress determined as per item 6.4.

6.2.2 DIMENSIONING BY THE ANDREE-FRICKE THEORY

The theory developed by Andree and Fricke for dimensioning of the wheel track is based on the assumption that the diagram of the pressure between the beam base and the concrete has the shape of a parabola. This theory was being applied for a long time with good results in the design of wheel tracks of gates, traveling cranes and other equipment.

The maximum value of the longitudinal bending stress of the wheel track is given by

$$\sigma_f = \frac{M}{W} = \frac{P}{2W} \sqrt[3]{\frac{E_a I}{E_c b}} \qquad (6.15)$$

where:

E_a and E_c = modulus of elasticity of the beam and of the foundation, respectively
I = moment of inertia of the cross section of wheel track
b = width of the wheel track
P = load transmitted by the wheel
W = section modulus.

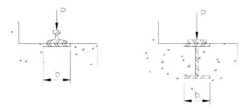

Fig. 6.7 Cross-section of wheel track

The bearing pressure between the wheel track base and the concrete is determined by the formula

$$p_o = \frac{P}{bd} \qquad\qquad (6.16)$$

where:

b = width of the wheel track
d = minimum value between L_1 and L_2
L_1 = spacing between two consecutive wheels
L_2 = fictitious length of the concrete area under compression, given by

$$L_2 = 3.55 \sqrt[3]{\frac{E_a I}{E_c b}} \qquad\qquad (6.17)$$

Example 6.1 A fixed-wheel gate with 20 wheels is being subjected to a hydraulic load of 24 MN. The wheels are spaced 500 mm apart and the wheel track comprises a steel plate with a cross section of 600 mm x 100 mm, as indicated:

$$I = 600 \cdot 100^3/12 = 50 \cdot 10^6 \text{ mm}^4$$
$$W = 600 \cdot 100^2/6 = 10^6 \text{ mm}^3$$

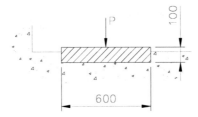

Fig. 6.8

Determine the maximum bending stress of the wheel track and the concrete bearing pressure. Assume: (a) the wheels are equally loaded; and (b) the second stage concrete has a characteristic compression strength (f_{ck}) of 22 MPa.

Solution:

a) According to the general theory of a beam on an elastic foundation:

- Load per wheel

$$P = 24/20 = 1.2 \text{ MN} = 1.2 \cdot 10^6 \text{ N}$$

-    Modulus of elasticity of concrete (E_c)

According to the Brazilian standard NBR-8883, the modulus of elasticity of the concrete can be determined by the formula:

$$E_c = 5940\sqrt{f_{ck}} + 3.5$$

where f_{ck} is given in MPa. In the present case, $f_{ck} = 22$ MPa. So,

$$E_c = 5940\sqrt{22+3.5} = 30000\,\text{MPa}$$

- Modulus of the foundation (Eq. 6.2c)

- β coefficient (Eq. 6.7)

$$k = 0.4 \cdot 30000 \sqrt[3]{\frac{30000 \cdot 600^4}{206000 \cdot 50 \cdot 10^6}} = 86725\,\text{MPa}$$

$$\beta = \sqrt[4]{\frac{86725}{4 \cdot 206000 \cdot 50 \cdot 10^6}} = 0.0068\,\text{mm}^{-1}$$

- Bending moment
First establish the origin of the coordinates in the contact point of the first wheel. Considering the overlapping effect only of the three first wheels, the bending moment under the first wheel, according to Equation 6.5a, will be:

$$M_1 = \frac{P}{4\beta}\left(C_{\beta x1} + C_{\beta x2} + C_{\beta x3}\right)$$

$\beta_{x1} - 0.0068 \cdot 0 = 0$ $\qquad \therefore C_{\beta x1} = e^0 (\cos 0 - \sin 0) = 1$

$\beta_{x2} = 0.0068 \cdot 500 = 3.39$ rad $\qquad \therefore C_{\beta x2} = e^{-3.39} (\cos 3.39 - \sin 3.39) = -0.0246$

$\beta_{x3} = 0.0068 \cdot 1000 = 6.77$ rad $\qquad \therefore C_{\beta x3} = e^{-6.77} (\cos 6.77 - \sin 6.77) = 0.0005$

$$M_1 = \frac{1.2 \cdot 10^6}{4 \cdot 0.0068}\left(1 - 0.0246 + 0.0005\right) = 43.216 \cdot 10^6\,\text{N-mm}$$

- Bending stress

$$\sigma_f = \frac{M}{W} = \pm\frac{43.216 \cdot 10^6}{10^6} = \pm 43.2\,\text{MPa}$$

- Concrete bearing pressure (Eq. 6.14)

$$p_o = \frac{P\beta}{2b}\left(A_{\beta x1} + A_{\beta x2} + A_{\beta x3}\right)$$

$\beta_{x1} = 0$ $\qquad \therefore A_{\beta x1} = e^0 (\cos 0 + \sin 0) = 1$

$\beta_{x2} = 3.39$ rad $\qquad \therefore A_{\beta x2} = e^{-3.39} (\cos 3.39 + \sin 3.39) = -0.0410$

$\beta_{x3} = 6.77$ rad $\qquad \therefore A_{\beta x3} = e^{-6.77} (\cos 6.77 + \sin 6.77) = 0.0015$

Therefore,

$$p_o = \frac{1.2 \cdot 10^6 \cdot 0.0068}{2 \cdot 600}\left(1 - 0.0410 + 0.0015\right) = 6.5\,\text{MPa}$$

b) By the Andree-Fricke method

- Bending stress (Eq. 6.15)

$$\sigma_f = \pm \frac{1.2 \cdot 10^6}{2 \cdot 10^6} \sqrt[3]{\frac{206000 \cdot 50 \cdot 10^6}{30000 \cdot 600}} = \pm 49.8 \text{MPa}$$

- Concrete bearing pressure

The wheels spacing is $L_1 = 500$ mm. According to Equation 6.17, the length L_2 is

$$L_2 = 3.55 \sqrt[3]{\frac{206000 \cdot 50 \cdot 10^6}{30000 \cdot 600}} = 295 \ \text{mm}$$

Since L_2 is smaller than L_1, then, by Equation 6.16,

$$p_o = \frac{1.2 \cdot 10^6}{600 \cdot 295} = 6.78 \text{ MPa}$$

Example 6.2 Suppose that the wheel track of the gate mentioned in the preceding example is now the one in Figure 6.9, and then calculate the bending stresses and the concrete bearing pressure.

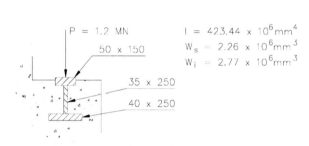

Fig. 6.9

Solution:

a) According to the general theory of the beam on elastic foundation

- Modulus of the foundation (Eq. 6.2c)

$$k = 0.4 \cdot 30000 \sqrt[3]{\frac{30000 \cdot 250^4}{206000 \cdot 423.44 \cdot 10^6}} = 13241 \text{MPa}$$

- β-coefficient (Eq. 6.7)

$$\beta = \sqrt[4]{\frac{13241}{4 \cdot 206000 \cdot 423.44 \cdot 10^6}} = 0.00248 \text{ mm}^{-1}$$

- Bending moment

The contact point of the first wheel will be taken as the origin of the coordinates. Considering the overlapping effect of only the three first wheels, gives:

$$M_1 = \frac{P}{4\beta}\left(C_{\beta x1} + C_{\beta x2} + C_{\beta x3}\right)$$

$\beta_{x1} = 0.00248 \cdot 0 = 0$
$\beta_{x2} = 0.00248 \cdot 500 = 1.241$
$\beta_{x3} = 0.00248 \cdot 1000 = 2.482$

The due calculations made,

$C_{\beta x1} = 1$ $A_{\beta x1} = 1$
$C_{\beta x2} = -0.1799$ $A_{\beta x2} = 0.3671$
$C_{\beta x3} = -0.1173$ $A_{\beta x3} = -0.0138$

Therefore,

$$M_1 = \frac{1.2 \cdot 10^6}{4 \cdot 0.00248}\left(1 - 0.1799 - 0.1173\right) = 84.948 \cdot 10^6 \text{ N-mm}$$

The bending stresses will be:

$\sigma_{fs} = -\left(84.948 \cdot 10^6 / 2.26 \cdot 10^6\right)$ $= -37.59$ MPa (compression)

$\sigma_{fi} = \left(84.948 \cdot 10^6 / 2.77 \cdot 10^6\right)$ $= 30.67$ MPa (tension)

- Concrete bearing pressure (Eq. 6.14)

With the values of $A_{\beta x}$ already determined,

$$p_o = \frac{1.2 \cdot 10^6 \cdot 0.00248}{2 \cdot 250}\left(1 + 0.3671 - 0.0148\right) = 8.06 \text{MPa}$$

b) By the Andree-Fricke method

- Bending stresses (Eq. 6.15)

$$\sigma_{fs} = -\frac{1.2 \cdot 10^6}{2 \cdot 2.26 \cdot 10^6}\sqrt[3]{\frac{206000 \cdot 423.44 \cdot 10^6}{30000 \cdot 250}} = -60.15 \text{ MPa}$$

$$\sigma_{fi} = 60.15\frac{2.26 \cdot 10^6}{2.77 \cdot 10^6} = 49.07 \cdot \text{MPa}$$

- Concrete bearing pressure

$L_1 = 500$mm

$$L_2 = 3.55 \sqrt[3]{\frac{206000 \cdot 423.44 \cdot 10^6}{30000 \cdot 250}} = 804 \text{mm}$$

Since L_2 is greater than L_1,

$$p_0 = \frac{P}{bL_1} = \frac{1.2 \cdot 10^6}{250 \cdot 500} = 9.6 \text{ MPa}$$

6.3 SLIDE TRACKS

The hydraulic load acting on the stoplogs or slide gates is transmitted by the vertical end girders to the slide track and from there to the foundation. When dimensioning the slide track, the concrete bearing pressure and the slide track bending stresses must be checked.
The slide track is usually designed as a beam on an elastic foundation, with finite length in the direction transversely to the flow and longitudinal unit length. Hetényi recommends, according to the loading case, the following equations:

a) Concentrated force at the middle

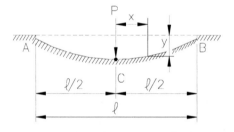

Fig. 6.10

Deflection at the middle is given by

$$y_c = \frac{P\beta}{2k} \frac{\text{Cosh } \beta 1 + \cos \beta 1 + 2}{\text{Sinh } \beta 1 + \sin \beta 1} \tag{6.18}$$

and at the end points,

$$y_A = y_B = \frac{2P\beta}{k} \frac{\text{Cosh } \dfrac{\beta 1}{2} \cos \dfrac{\beta 1}{2}}{\text{Sinh } \beta 1 + \sin \beta 1} \tag{6.19}$$

The deflection at the end points is equal to zero for $\beta l = \pi$, 3π, 5π and so on, and positive when βl is less than π. The expression

$$l_{ef} = \pi/\beta \tag{6.20}$$

defines the effective length of the beam (see Figure 6.11).

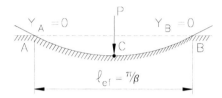

Fig. 6.11

In the above expressions, P is the load per unit length and l the width of the slide track. The section modulus is given by

$$W = bt^2/6 = t^2/6 \tag{6.21}$$

as $b = 1$ (unit length); t is the slide track thickness.
The maximum bending moment occurs under the load and is determined by

$$M_C = \frac{P}{4\beta}\frac{\cosh\beta l - \cos\beta l}{\sinh\beta l + \sin\beta l} \tag{6.22}$$

Hence, the bending stress is

$$\sigma_f = M_c/W = 6\,M_c/t^2 \tag{6.23}$$

The concrete bearing pressure under the slide track, considering an uniform pressure distribution, is calculated by

$$p_o = \frac{P}{b\,l_{ef}} = \frac{P}{l_{ef}} \tag{6.24}$$

as $b = 1$.
In cases where l_{ef} results in a value that is greater than the actual width l, the former should be used in the above expression.

b) Concentrated force at an arbitrary point

The equations deduced by Hetényi to determine the deflection and the bending moment at any point of the beam will be presented. The formulas are for the A-C portion of the beam, where $x < a$. The same formulas can be used for the B-C section, where $x < b$, by measuring x from end B and replacing a by b and b by a.

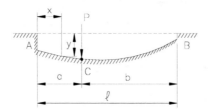

Fig. 6.12

The deflection at any point of A-C portion is

$$y = \frac{P\beta}{k} \frac{1}{Sinh^2\beta l - sin^2\beta l} \{2Cosh\beta x \cos\beta x (Sinh\,\beta l\cos\beta a\,Cosh\,\beta b - sin\,\beta l\,Cosh\,\beta a\,\cos\beta b)$$

$$+ (Cosh\,\beta x\,sin\beta x + Sinh\,\beta x\,\cos\beta x) \left[Sinh\,\beta l (sin\,\beta a\,Cosh\,\beta b - \cos\beta a\,Sinh\,\beta b) + \right.$$
$$\left. + sin\,\beta l (Sinh\,\beta a\,\cos\beta b - Cosh\,\beta a\,sin\beta b) \right] \} \tag{6.25}$$

At the point of application of the load ($x = a$), the deflection is

$$y_C = \frac{P\beta}{k} \frac{1}{Sinh^2\beta l - sin^2\beta l} \left[\left(Cosh^2\,\beta a + cos^2\,\beta a \right)(Sinh\,\beta b\,Cosh\,\beta b - sin\,\beta b\,cos\beta b) + \right.$$

$$\left. + \left(Cosh^2\,\beta b + cos^2\,\beta b \right)(Sinh\,\beta a\,Cosh\,\beta a - sin\,\beta a\,cos\beta a) \right] \tag{6.26}$$

The bending moment in the A-C section is

$$M_x = \frac{P}{2\beta} \frac{1}{Sinh^2\beta l - sin^2\beta l} \{2\,Sinh\,\beta x\,sin\,\beta x (Sinh\,\beta l\cos\beta a\,Cosh\,\beta b - sin\beta l\,Cosh\,\beta a\,\cos\beta b)$$

$$+ (Cosh\,\beta x\,sin\,\beta x - Sinh\,\beta x\,\cos\beta x)\left[Sinh\,\beta l (sin\,\beta a\,Cosh\,\beta b - \cos\beta a\,Sinh\,\beta b) + \right.$$

$$\left. + sin\,\beta l (Sinh\,\beta a\,\cos\beta b - Cosh\,\beta a\,sin\beta b) \right] \} \tag{6.27}$$

Bending moment at the point of application of the load:

$$M_c = \frac{P}{4\beta} \frac{1}{Sinh^2\beta l - sin^2\beta l} \left[\left(Cosh^2\,\beta a - cos^2\,\beta a \right)(Sinh\,2\beta b - sin\,2\,\beta\,b) + \right.$$

$$\left. + \left(Cosh^2\,\beta b - cos^2\,\beta b \right)(Sinh\,2\beta a - sin\,2\beta a) \right] \tag{6.28}$$

Example 6.3 A 2 m high stoplog is subjected to a hydraulic load of W = 5400 kN. Design the slide track, knowing that E_c is 30000 MPa and the slide track width is 220 mm.

Solution:

Suppose that the stoplog vertical end girders act on the middle of the slide track, as per Figure 6.13.

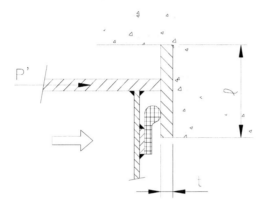

Fig 6 13

- Load on the slide track

The load transmitted by each vertical end girder is

$\quad$ P' = W/2 = 5400/2 = 2700 kN.

The height of the vertical end girder is 2 m. Therefore, the mean load value per unit length will be

$\quad$ P = 2700/2000 = 1.35 kN = 1350 N.

- Modulus of the foundation

We have:
$\quad$ E_c = 30000 MPa
$\quad$ E_a = 206000 MPa
$\quad$ b = 1 mm (unit length)
$\quad$ $I = bt^3/12$
$\quad$ $W = bt^2/6$

Consider that the thickness of the slide track is $t = 40$ mm. So,

$$I = 1 \cdot 40^3/12 = 5333 \text{ mm}^4, \text{ and } W = 1 \cdot 40^2/6 = 266.7 \text{ mm}^3$$

According to Equation 6.2c, the modulus of the foundation will be

$$k = 0.4 \cdot 30000 \sqrt[3]{\frac{30000 \cdot 1^4}{206000 \cdot 5333}} = 361.36 \text{ MPa}$$

- β coefficient (Eq. 6.7)

$$\beta = \sqrt[4]{\frac{361.36}{4 \cdot 206000 \cdot 5333}} = 0.0169 \text{ mm}^{-1}$$

- Effective length (Eq. 6.20)

$$l_{ef} = \pi/\beta = \pi/0.0169 = 185.5 \text{ mm} < 220 \text{ mm}.$$

- Maximum bending moment (Eq. 6.22):

$\beta l = 0.0169 \cdot 220 = 3.725 \text{ rad}$
$\text{Cosh } 3.725 = \frac{1}{2} (e^{3.725} + e^{-3.725}) = 20.758$
$\text{Sinh } 3.725 = \frac{1}{2} (e^{3.725} - e^{-3.725}) = 20.734$
$\cos 3.725 = -0.8343$
$\sin 3.725 = -0.5513$

Therefore,

$$M_c = \frac{1350}{4 \cdot 0.0169} \frac{20.759 - (-0.8343)}{20.734 + (-0.5513)} = 21366 \text{ N-mm}$$

- Bending stress

$$\sigma = M_c/W = 21366/266.7 = 80.1 \text{ MPa}$$

- Concrete bearing pressure (Eq. 6.24)

$$p_o = 1350/185.5 = 7.28 \text{ MPa}.$$

6.4 CONCRETE BEARING PRESSURE

The mean concrete bearing pressure calculated according to Equations 6.14, 6.16 or 6.24 should not exceed the comparative stress, f_c, given by:

a) in the case where the concrete immediately below the transmission surface is not reinforced:

$$f_c = 0.325\, f_{ck} \qquad\qquad \text{for } f_{ck} \le 18 \text{ MPa} \qquad\qquad (6.29)$$

$$f_c = 0.195\, f_{ck} + 2.31 \qquad \text{for } f_{ck} > 18 \text{ MPa} \qquad\qquad (6.30)$$

b) in the case where the concrete immediately below the transmission surface is reinforced (see Figure 6.14):

$$f_c = \frac{f_{ck}}{2.1} \sqrt[3]{\frac{b^*}{b}} \le 15\,\text{MPa} \qquad\qquad (6.31)$$

where:
   $b^* = 3b$, in the case of $e \ge b$
   $b^* = b + 2e$, in the case of $e < b$
   f_{ck} = characteristic compression strength of concrete, in MPa
   b^* and b = dimensions as per Figure 6.14
   e = distance between the concrete outer face and the extreme edge of the wheel track beam base.

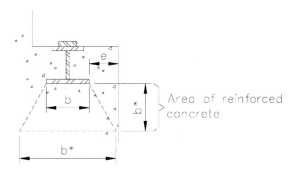

Fig. 6.14

With the exception of guide rails, the minimum distance of the outer edge of the rail or the rail girder from the free concrete surface should be $e = 120$ mm. It may be reduced, if steel plate cladding (armored plating) is installed on the concrete surface, and if this cladding is sufficiently anchored by reinforcement bars transversely installed.

When the first version of the Brazilian standard for gate calculation NBR-8883 was being developed, it was observed that the DIN 19704 standard, in its December 1963 edition, considered 'a priori' the concrete below the transmission surface as being reinforced (see item 5.5.5 of said standard). The September 1976 edition of the same standard determines that the concrete bearing pressure should be calculated according to the DIN 1045 standard, January 1972 edition, item 17.3.3, which is concerned exclusively with reinforced concrete. Now, this situation does not occur always, since wheel or slide tracks are usually installed in slots, which later on are filled with non-reinforced second stage concrete. Therefore, it is necessary to separate the cases of reinforced and non-reinforced concrete and determine the corresponding allowable bearing stress. The following will show how Equations 6.29, 6.30 and 6.31 have been determined.

Case of Non-Reinforced Concrete

Given a continuous plate, of width b, supported on a plain concrete structure and subjected to a single load P. This load gives rise to compression and tension stresses along the x-axis, as shown in Figure 6.15. These stresses depend on the bearing stress σ_c between the plate and the concrete and on the ratio d/b (d is the width of the concrete block and b is the plate width).

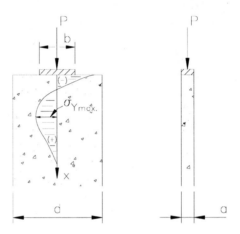

Fig. 6.15 Stress distribution along x-axis

According to the results of laboratory tests referred by Leonhardt and Monnig [5], the critical situation occurs for $d/b = 2$, when

$$\sigma_{ymax} = 0.11\,\sigma_c \tag{6.32}$$
where
$$\sigma_c = P/ab \tag{6.33}$$
$\sigma_{y\,max}$ = maximum tensile stress.

The tensile stress, in turn, has as limit value, when re-bars against cracking are not used:

$$1.4\,\sigma_{y\,max} \le f_{tk}/2 \tag{6.34}$$

where
f_{tk} is the characteristic tensile strength of concrete

or,
$$\sigma_{y\,max} \le f_{tk}/2.8 \tag{6.35}$$

Substituting $\sigma_{y\,max}$ in Equation 6.32,

$$\sigma_c = \frac{\sigma_{ymax}}{0.11} \le \frac{f_{tk}}{0.11 \cdot 2.8}$$

Therefore,

$$\sigma_c = \leq 3.247 f_{tk} \tag{6.36}$$

Item 5.2.1.2 of the Brazilian standard NB-1 on concrete, states:

- for $f_{ck} \leq 18$ MPa,

$$f_{tk} = (1/10) f_{ck} \tag{6.37}$$

- for $f_{ck} > 18$ MPa,

$$f_{tk} = 0.06 f_{ck} + 0.71 \tag{6.38}$$
where
f_{ck} is the characteristic compression strength of concrete.

So,

- for $f_{ck} \leq 18$ MPa,

$$\sigma_c \leq 3.247 f_{tk} = 3.247 \cdot (1/10) f_{ck}$$
$$\therefore \sigma_c \leq 0.325 f_{ck} \tag{6.39}$$

- for $f_{ck} > 18$ MPa,

$$\sigma_c \leq 3.247 f_{tk} = 3.247 (0.06 f_{ck} + 0.71)$$
$$\therefore \sigma_c \leq 0.195 f_{ck} + 2.31 \tag{6.40}$$

The two last equations correspond, respectively, to Equations 6.29 and 6.30.

Case of Reinforced Concrete

The Brazilian standard on concrete NB-1 says, in item 5.3.1.2.e (quote):

'5.3.1.2 - Ultimate values of the design stresses

e) Bearing pressures in reduced areas

In members loaded in a reduced area A_o on one of the faces and with a height not less than the largest width, and in members loaded in a reduced area A_o on two opposite faces and with a height not less than twice the largest width, the ultimate value of the design stress is

$$f_{cd} \sqrt[3]{\frac{A_c}{A_o}} \leq 21 \text{ MPa} \tag{6.41}$$

taking for A_c the area of the geometric figure which, having the same center of gravity as A_o, is the maximum one the member surface can be included.'
Note: f_{cd} is the design compression strength of concrete.

Nevertheless, the NB-1 standard operates with ultimate values of the design stress and with the load multiplied by a safety factor ν_f. This factor can be considered equal to 1.4, according to item 5.4.2.1 - *'Ultimate Limit State'* of NB-1.

So, the expression 6.41 may be rewritten:

$$\gamma_f \, \sigma_c \leq f_{cd} \sqrt[3]{\frac{A_c}{A_o}} \leq 21\,\text{MPa} \tag{6.42}$$

The relation between f_{cd} and f_{ck} is given by the item 5.4.1 of NB-1 and is

$$f_{cd} = f_{cd}/\gamma_c \tag{6.43}$$

where:

γ_c is the strength reduction coefficient, in this case equal to 1.5.

Thus,

$$\sigma_c \leq \frac{f_{cd}}{\gamma_f} \sqrt[3]{\frac{A_c}{A_o}} = \frac{f_{ck}}{\gamma_f \, \gamma_c} \sqrt[3]{\frac{A_c}{A_o}} = \frac{f_{ck}}{1.4 \cdot 1.5} \sqrt[3]{\frac{A_c}{A_o}}$$

$$\therefore \sigma_c \leq \frac{f_{ck}}{2.1} \sqrt[3]{\frac{A_c}{A_o}} \tag{6.44}$$

As Equation 6.42 gives as limit

$$\gamma_f \, \sigma_c \leq 21 \text{ MPa},$$

Thus,

$$\sigma_c \leq 21/1.4 = 15 \text{ MPa} \tag{6.45}$$

In case of continuous plate supported on the concrete,

$$A_c/A_o = b^*/b$$

and the final equation becomes

$$\sigma_c \leq f_{ck} \sqrt[3]{\frac{b^*}{b}} \leq 15\,\text{MPa} \tag{6.46}$$

which corresponds to Equation 6.31.

6.5 LATERAL GUIDANCE

When totally open, the flat gate should maintain a vertical line in the plane normal to the flow, because once the gate starts to descend and cross the hydraulic flow, the moving direction stays unaltered due to the great force required to make the wheels slide laterally on the wheel track. Tests show that the force in the axial direction reaches up to about 30 per cent of the radial load [6]. It is interesting to note that the gate weight is small when compared with the water thrust on the gate and, therefore, a very large force is necessary to restore the gate to the vertical position. So, if the gate has an inclined position in its plane when starting to close, one of the corners of the skin plate bottom edge will press against the sill beam and its full closure will not be attained. Also, if the main wheels are flanged and the flanges press against the wheel track, a gate jamming may occur in a partially open position. Gates with lateral guidance through rigid shoes may be subjected to the same problem.

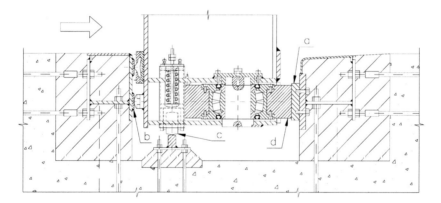

Fig. 6.16 Spring loaded side roller of the intake gate at the Portage Mountain Power Plant
(a) wheel track; (b) counterguide; (c) side roller; (d) main wheel

The most adequate method to keep the gate vertical consists in the use of auxiliary wheels mounted on elastic supports (springs or rubber pads), as shown in Figure 6.16.
In this system, the guide wheels are kept permanently in contact with the lateral embedded parts and the gate is centered by the spring action. The limitation of the gate lateral displacement through shoes or flanged main wheels should be restricted to small gates.
   In case of wheels mounted on cylindrical self-lubricating bushings, suspending the wheels between 'Belleville' conical disc springs eliminates the danger of overloading by excessive lateral forces. In this arrangement, the wheel pins are supported at both ends. When the gate is raised, the wheels are kept centered by the springs. Depending on the design, forces acting on Belleville disc springs remain fairly constant irrespective of the spring deflection [7, 8]. Due to this feature, with the gate moving under water load, the wheels can travel laterally and the axial forces transmitted to the vertical end girders are limited.

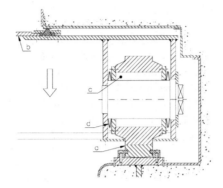

Fig. 6.17 Conical disc springs for wheel centering
(a) wheel; (b) skin plate; (c) bushing; (d) conical disc spring

6.6 WHEELS AND PINS

6.6.1 DESIGN FEATURES

The main wheels are designed to support the maximum water thrust acting on the gate. For economical reasons, it is recommended to determine their spacing so that all of them receive the same portion of the water thrust (see item 4.2.3). Wheels are mounted either on roller bearings, grease lubricated or self-lubricating bronze bushings. In the case of cylindrical bushings, the wheel tread face should be crowned, in order to follow the pin inclination caused by deflection of the gate horizontal beams, thus avoiding stress concentration near the inner edge of the wheel track. The curvature radius of the wheel tread face is selected in the range of 10 to 15 times the wheel radius.

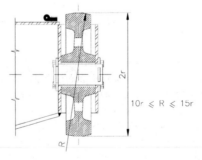

Fig. 6.18 Wheel crown radius

For wheels mounted on self-aligning roller bearings, the rolling surface may be cylindrical, for these bearings allow small angular displacement of the pin. When roller bearings are selected, the manufacturer's specifications of limit values of angular misalignment between the inner and outer rings should be followed.

The horizontal beams deflect under load and tilt the vertical end girders and wheel pins mounted therein. The elastic angle deflection at the ends of horizontal beams with constant cross-section is given, in radians, by

$$\theta = \frac{F L^2}{24 E I}$$

(6.47)

where

F = water thrust on the horizontal beam
L = support length
E = modulus of elasticity
I = moment of inertia of the beam cross-section.

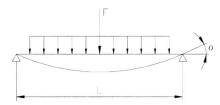

Fig. 6.19 Deflection of the horizontal girder

Example 6.4 Determine the slope of the wheel pins of a fixed-wheel gate, whose horizontal beams are subjected to an unit water thrust of 970 kN. The wheels support span is 4500 mm and the horizontal beams cross-section has a moment of inertia of $1.55 \cdot 10^9$ mm^4.

Solution:

$$\theta = \frac{970000 \cdot 4500^2}{24 \cdot 206000 \cdot 1.55 \cdot 10^9} = 2.56 \cdot 10^{-3} \text{rad} = \quad 0.15^0$$

Table 6.2 shows the allowable values of angular misalignments for self-aligning roller bearings manufactured by SKF, according to the dimension series.

Table 6.2 Maximum Permissible Angular Misalignment of SKF Self-Aligning Roller Bearings

Bearing dimension series	Maximum permissible angular misalignment (Degrees)
213	1.0
222C, 222	1.5
223C	2.0
230C	1.5
231C	1.5
232C	2.5
238C	1.0
239C	1.5
240C	2.0
241C	2.5

It should be noted that the use of two self-aligning roller bearings in the same wheel prevents angular displacement of the wheel in relation to the pin. In this case, the wheel tread face should be crowned.

To prevent weakening of wheels mounted on roller bearings, the outside bearing diameter should not exceed 80 per cent that of the wheel.

The pins may be mounted either in cantilever (Fig. 6.20) or supported at the ends (Fig. 6.4). Cantilever-mounted wheels require fairly large wheel pins, but provide a maximum of accessibility for inspection and maintenance of wheels and bearings. As a rule, the pins are eccentrically turned in the lathe to allow proper alignment of the wheel tread faces in the final assembly. A 3 mm eccentricity is usually enough.

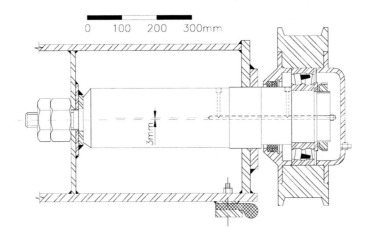

Fig. 6.20 Fixed-wheel gate, Passo Real Power Plant, 2.5 m wide by 5 m high

After final alignment of the wheels tread faces, the pins are prevented turning by means of key plates or nuts and lock nuts. In gates intended to remain submerged for long periods of time, special attention should be given to the bearing and wheel cover seals so as to prevent the entrance of water and sediment, which might damage the bearings. Seals mounted in turning parts should make contact with surfaces resistant to corrosion.

The use of grease fittings is recommended, even in the case of self-lubricating bushings, for the grease deposited within the wheel fills the empty spaces and prevents the entrance of water or debris.

6.6.2 CONTACT PRESSURE BETWEEN WHEEL AND TRACK

The design of wheels is made as a function of the contact pressure between wheel and track. Depending on the geometry of the wheel tread face and the wheel track, the contact is rectangular or elliptical. The rectangular contact takes place exclusively on cylindrical wheels and a flat wheel track. If the wheel has a crowned tread face or the wheel track is curved in the horizontal plane, the contact is elliptical.

When designing the wheels, the first step is the choice of the geometry and the basic dimensions of the wheel tread face and wheel track. The contact pressure is then determined from the corresponding formula. As a starting point for the choice of the wheel diameter, the following empirical relation may be used:

$$18 \sqrt{P} \langle D \langle 22.5 \sqrt{P}$$

(6.48)

where:

D = wheel diameter, in mm
P = wheel load, in kN.

The width of the wheel tread face may be taken as one forth to one sixth of the wheel diameter, for preliminary calculations.

If the calculated contact pressure exceeds the allowable pressure, the wheel diameter (and the width, if necessary) is increased and the pressure recalculated, until the required result is attained. From an economic viewpoint, it is not interesting to keep the calculated contact pressure much below the allowable one, for the wheel would be oversized.

a) Case of rectangular contact

The Hertz formula is used for finding the contact pressure between wheel and track, in case of rectangular contact:

$$\sigma_c = 0.418 \sqrt{\frac{PE}{R\,1}}$$

(6.49)

where:

P = wheel load
E = modulus of elasticity of steel
R = wheel tread radius
1 = wheel tread width.

It should be noted that the Hertz theory gives the maximum compressive stresses that occur at the center of the surfaces of contact of a cylinder and a flat plate but not the maximum shear stresses, which occur in the interior of the compressed parts. The theory assumes the length of the cylinder and dimensions of the plate to be infinite. For a very short cylinder and for a plate having a width less than five to six times that of the contact area or a thickness less than five to six times the depth to the point of maximum shear stress, the actual stress may vary considerably from the values indicated by the theory [9].

The maximum shear stress is given by

$$\tau_{max} \approx 1/3\ \sigma_c$$

(6.49a)

and occurs at a depth of $h' = 0.4\ b$ below the surface of the plane, where

$$\sigma_c = 3.041 \sqrt{\frac{PR}{1E}}$$

(6.49b)

The minimum thickness of the wheel track (or the depth of the weld root opening of the wheel track butt joints) should be at least five to six times 0.4 b, that is, from 2 b up to 2.4 b.

Example 6.5 Calculate the contact pressure between the wheels and tracks of a gate subjected to a water thrust of 4800 kN. The gate has eight wheels with a diameter of 450 mm and a 80 mm wide cylindrical wheel tread face. The wheels are equally loaded and the wheel track is flat. Also, determine the maximum shear stress and the minimum thickness of the wheel track.

Solution:

The contact between the wheels and the wheel track is rectangular.

- Wheel load
    P = 4800/8 = 600 kN
    E = 206000 MPa
    R = 450/2 = 225 mm
    l = 80 mm

- Contact pressure:

$$\sigma_c = 0.418 \sqrt{\frac{600000 \cdot 206000}{225 \cdot 80}} = 1095 \text{ MPa}$$

- Maximum shear stress:

$$\tau_{max} \approx 1/3 \, \sigma_c = 1/3 \cdot 1095 = 365 \text{ MPa}$$

It occurs at a depth of $h' = 0.4 \, b$, where

$$b = 3.041 \sqrt{\frac{600000 \cdot 225}{80 \cdot 206000}} = 8.7 \text{ mm}$$

$h' = 0.4 \, b = 0.4 \cdot 8.7 = 3.48$ mm.

- Minimum thickness of wheel track:

$$5 \, h' = 5 \cdot 3.48 = 17.41 \text{ mm.}$$

b) Case of elliptical contact

In the case of elliptical contact, the maximum contact pressure occurs in the ellipse center and, according to Timoshenko [10], has the value of

$$\sigma_c = \frac{3 \, P}{2 \pi a b} \tag{6.50}$$

where:

P = wheel load

a, b = semi-axes of the ellipse of contact.

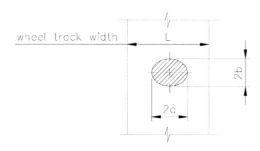

Fig. 6.21 Ellipse of contact

The semi-axes of the ellipse are determined by

$$a = \alpha \sqrt[3]{\frac{P\,m}{n}} \qquad\qquad (6.51)$$

and

$$b = \beta \sqrt[3]{\frac{P\,m}{n}} \qquad\qquad (6.52)$$

where

$$m = \frac{4}{\dfrac{1}{r_1} + \dfrac{1}{r'_1} + \dfrac{1}{r_2} + \dfrac{1}{r'_2}} \qquad\qquad (6.53)$$

$$n = \frac{4\,E}{3\left(1 - \mu^2\right)} \qquad\qquad (6.54)$$

E = modulus of elasticity of steel

μ = Poisson ratio

r_1 and r'_1 = main curvature radii of the wheel, in the contact point

r_2 and r'_2 = main curvature radii of the wheel track, in the contact point.

The α and β-coefficients indicated by Timoshenko are taken from Table 6.3, as a function of the θ angle, which is given by

$$\theta = \text{arc cos } B/A \qquad (6.55)$$

where:

$$A = 2/m \qquad (6.56)$$

$$B = \frac{1}{2}\sqrt{\left(\frac{1}{r_1} - \frac{1}{r'_1}\right)^2 + \left(\frac{1}{r_2} + \frac{1}{r'_2}\right)^2 + 2\left(\frac{1}{r_1} - \frac{1}{r'_1}\right)\left(\frac{1}{r_2} - \frac{1}{r'_2}\right)\cos 2\rho} \qquad (6.57)$$

and ρ is the angle between the plans containing the main curves $1/r_1$ and $1/r_2$. It is important to note that in the case of a flat wheel track,

$$r_2 = \infty, \qquad \text{and } r'_2 = \infty$$

Therefore,

$$1/r_2 = 0 \qquad \text{and } 1/r'_2 = 0$$

So, Equations 6.53 and 6.57 are reduced to

$$m = \frac{4}{\dfrac{1}{r_1} + \dfrac{1}{r'_1}} \qquad (6.53a)$$

and

$$B = \frac{1}{2}\left(\frac{1}{r_1} - \frac{1}{r'_1}\right) \qquad (6.57a)$$

Table 6.3 Coefficients α and β (Timoshenko)

θ (degrees)	α	β	θ (degrees)	α	β
20	3.778	0.408	60	1.486	0.717
30	2.731	0.493	65	1.378	0.759
35	2.397	0.530	70	1.284	0.802
40	2.136	0.567	75	1.202	0.846
45	1.926	0.604	80	1.128	0.893
50	1.754	0.641	85	1.061	0.944
55	1.611	0.678	90	1.000	1.000

The data shown in Table 6.3 are plotted in Figure 6.22.

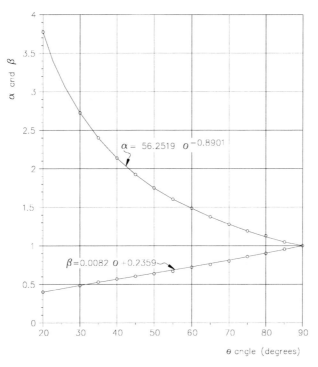

Fig. 6.22 Coefficients α and β (Timoshenko)

The wheel tread width is determined so that under maximum gate deflection there is a distance of about 10 mm between the contact ellipse and the track edge. The final wheel track width is established so as to accommodate eventual side displacement of the gate.

Example 6.6 Assume that the gate wheels of the preceding example have a crowned tread face with a 2700 mm radius, and recalculate the contact pressure.

Solution:

In this case, the contact between the wheels and wheel track has an elliptical shape and the pressure is calculated according to Equation 6.50.

- Wheel load
    $P = 600000$ N

- Semi-axes of the ellipse of contact

Thus,
    $r_1 = 450/2 = 225$ mm
    r'_1 = wheel tread crown radius = 2700 mm
    r_2 and r'_2 = wheel track curvature radii = ∞ (flat surface)

ρ = angle between the plans of the curves $1/r_1$ and $1/r_2 = 0$
E = modulus of elasticity of steel = 206000 MPa
μ = Poisson ratio = 0.3

Equations 6.53a and 6.54 give, respectively,

$$m = \frac{4}{\dfrac{1}{225} + \dfrac{1}{2700}} = 831\,mm$$

$$n = \frac{4 \cdot 206000}{3\left(1 - 0.3^2\right)} = 301831\ N/mm^2$$

Bringing these values to Equations 6.56 and 6.57a,

A = 2/831 = 0.00241
B = ½ (1/225 – 1/2700) = 0.00204
$\therefore \theta$ = arc cos (0.00204/0.00241) = 32.17°

Inserting this value in Table 6.3,

α = 2.586, and
β = 0.509.
So,

$$a = 2.586\ \sqrt[3]{\frac{600000 \cdot 831}{301831}} = 30.57\ mm$$

$$b = 0.509\ \sqrt[3]{\frac{600000 \cdot 831}{301831}} = 6.02\ mm$$

Therefore, the maximum contact pressure is

$$p_c = \frac{3 \cdot 600000}{2\pi \cdot 30.57 \cdot 6.02} = 1557\ MPa$$

6.6.3 PERMISSIBLE CONTACT STRESSES

The allowable contact stresses between wheels and tracks are determined as a function of the ultimate strength of the material and the frequency of movement of the gate, as set forth in Table 6.4. The values listed refer to unhardened contact surfaces and to normal loading case.

Table 6.4 Permissible Pressures between Wheel and Track (NBR-8883)

Contact surfaces	Frequency of movement	
	Low (less than 100/year)	High (more than 100/year)
A cylinder and a flat plate - e.g., wheel and wheel track	$1.85\,f_s$	$1.6\,f_s$
Two cylinders - e.g., axle and wheel (or roller)	$2\,f_s$	$1.9\,f_s$

The values of f_s (ultimate strength of steel) should always be referred to the weakest material (wheel or track). The allowable stresses listed in Table 6.4 are increased or decreased as follows:

a) in case of occasional loading, increase the values by 12 per cent;
b) in case of exceptional loading, it is not necessary to check the compression stresses;
c) for crowned contact surfaces with a radius ratio less or equal to 15:1, increase the values by 50 per cent;
d) for hardened surfaces, the stresses may be increased according to the hardness of the material;
e) for wheels and rollers temporarily immersed in water, the values of high frequency of movement should be used;
f) for wheels and rollers permanently immersed in water and exposed to large water flows, the tabulated values should be reduced as follows:

Motions per year	Reduction
up to 300	10 per cent
above 300 to 2000	15 per cent
above 2000 to 20000	30 per cent
above 20000	40 per cent

Example 6.7 Determine the allowable compression stress between wheel and track of a gate moved 200 times a year, knowing that:
a) the wheel track is flat and the wheel tread is crowned, with a radii ratio equal to 12:1;
b) the wheels are made of cast steel ASTM A27 Grade 70-40 (f_s = 480MPa) and the wheel track of steel SAE 1045 (f_s = 660 MPa). The contact surfaces are unhardened.

Solution:

Since the wheel track material is more resistant than that of the wheel, the allowable stress should be determined according to the ultimate strength of the A27 Grade 70-40 steel. For frequently moved equipment and contact between a flat and a curved surface, Table 6.4 gives

$$\sigma_{adm} = 1.6 f_s$$

As the radii ratio is 12:1 (less than 15:1), the allowable stress may be increased by 50 per cent. Thus,

- normal load case:

$$\sigma_a = 1.5 \cdot 1.6 \cdot 480 = 1152 \text{ MPa}$$

- occasional load case:
$$\sigma_a = 1.12 \cdot 1.5 \cdot 1.6 \cdot 480 = 1290 \text{ MPa.}$$

6.6.4 SURFACE HARDNESS

The wheel tread is usually manufactured with a hardness below that of the wheel track (about 50 Brinell). This is because replacement of the wheel track, if worn out, is impossible in most cases, whereas the wheels are easily inspected and repaired, or even replaced, if necessary. The wheel tread hardness is usually set at about 250 Brinell.

This requirement does not necessarily apply to temporary use equipment or to elements of lateral guidance or counterguidance.

6.7 BUSHINGS

Cylindrical bushings should be dimensioned based on the pressure given by

$$\sigma_d = \frac{P}{ld} \tag{6.58}$$

where:
  P = load on the bushing
  l = effective length of the bushing
  d = inner diameter of the bushing.

The calculated pressure should not exceed the values:

. for bronze bushings,   15 MPa
. for self-lubricating bushings,   35 MPa

The bushing inside diameter is determined from the pin diameter and its length should not be less than one and half times the inside diameter. The minimum recommended wall thickness is shown in Table 6.5.

Table 6.5 Bearing Wall Thickness

Inner diameter (mm)	Wall thickness (mm)
Up to 40	5
From 40 to 50	6
From 50 to 65	7
From 65 to 75	8
From 75 to 100	10
From 100 to 125	12

For sizes larger than those tabulated, use the formula:

$$t = \frac{d}{16} + 4 \tag{6.59}$$

where d is the bushing inside diameter (in mm).

REFERENCES

1. Ethembabaoglu, S.: Some Characteristics of Static Pressures in the Vicinity of Slots, *XIII ICOLD Congress*, Q50, R20 (1979), New Delhi.
2. Sharma, H.R.: Problems at High Head Gates in Outlet Conduits, *XIII ICOLD Congress*, Q50, R47 (1979), New Delhi.
3. Biot, M.A.: Bending of an Infinite Beam on an Elastic Foundation, ASME, *Journal of Applied Mechanics* (March 1937).
4. Hetényi, M.: *Beams on Elastic Foundation*, The University of Michigan Press, 7th printing (1964).
5. Leonhardt, F. and Monnig, E.: *Concrete Construction* (in Portuguese), Vol. 2.
6. Josserand, A. and Delaroche, J.: Les Vannes Aval Neyrpic, *Revue Technique NEYRPIC*, No.1 (1982).
7. Heckel, R.: Austrian Contributions toward the Technical Development of Hydraulic Steel Structures, *Stahlbau und Rundschau*, No. 26.
8. Kent's: *Mechanical Engineering Handbook*, John Wiley and Sons, New York, 12th edition (1964).
9. Young, W.C.: *Roark's Formulas for Stress and Strain*, 6th edition, McGraw-Hill Book Co. (1989).
10. Timoshenko, S.P.: *Strength of Materials* (in Portuguese), Vol. 2, Livros Técnicos e Científicos Editora (1981), Rio de Janeiro.

Chapter 7

Estimating Gate Weights

7.1 INTRODUCTION

In the early stages of a project, the type of gate is selected and the costs of its manufacture, transportation and erection are estimated. The gate weight directly affects the capacity of the hoisting equipment, and the final cost of supply is generally estimated on the basis of the total weight of the equipment, of which the weight of the gate is the biggest portion.

A precise determination of the weight of the gate is a rather complex task, which requires involved calculations. Generally speaking, it is necessary to calculate the hydraulic forces, loading on members, working stresses, bolted and welded connections, pivots and supports (bearings, trunnions, wheels, pins, bushings, and so on), and finally, the gate weight. All these calculations follow a slow trial-and-error process, in which the main geometric characteristics of members are first estimated and then the working stresses are calculated. Should a reasonable result not be reached, new characteristics are established and the calculations are repeated. However, in the initial phases of a project, there are not enough funds or time available to perform a comprehensive study, and it is usual to resort to comparisons with similar gates recently supplied or to graphical or analytical methods existing in the literature, such as those proposed by Schreiber [1], Davis and Sorensen [2], Boissonnault [3] and Navarro and Aracil [4].

In this chapter, equations are given for weight estimation of the following types of gates: spillway and submerged segment gates; fixed-wheel gates; double-leaf fixed-wheel gates; spillway and submerged stoplogs; flap gates; and caterpillar gates. For each gate type, an expression involving only the nominal gate dimensions and the head on the sill is developed, since these are generally the only defined characteristics at the beginning of the project.

The expressions for gate weight were determined by curve fitting of statistical data of 266 gates (154 installed in Brazil) listed in Tables 7.2 to 7.10, and are shown in Table 7.11 together with the respective coefficients of determination. These coefficients enable the degree of fitness to be evaluated. The expressions give only the weight of the moving part of the gate.

For all types except flap gates, the weight is given as a function of the parameter $B^2 hH$, were B and h are the span and the gate height respectively, and H is the head on the sill. In the equations, B, h and H are expressed in meters and the weight G is given in kN. For spillway gates, it can be assumed $H = h$, without incurring any substantial error. For flap gates the parameter $B^2 hH$ did not give consistent results, as these gates are not supported in the direction of their span. Another parameter $B(hH)^n$ was used, with satisfactory results.

It must be remembered that the expressions are based on statistical data and they are not recommended for precise estimates of gate weight. Accurate weight estimates can only be obtained by dimensioning all gate elements carefully, taking into consideration the particular conditions of the project, such as the available steel, and their characteristics, standards of design, allowable stresses and deflections, space limitations, concrete bearing capacity, corrosion allowance, and so on.

7.2 SEGMENT GATES

The weight of segment gates can be estimated by the equation

$$G = 0.64 \, (B^2 hH)^{0.682} \qquad (7.1)$$

for spillway gates, and

$$G = 3.688 \, (B^2 hH)^{0.521} \qquad (7.2)$$

for submerged gates. The weights of a selection of these gates are shown in the graphs of Figures 7.1 and 7.2, respectively.

Most of the spillway segment gates considered here have a pair of sloped arms at each side and their skin plate radius is equal to, or slightly larger than, the height. The dimensions and weight of the Furnas gates shown in Table 7.2 refer to the original design, after which they were heightened and reinforced to permit an increase in the reservoir water level of 1.5 m.

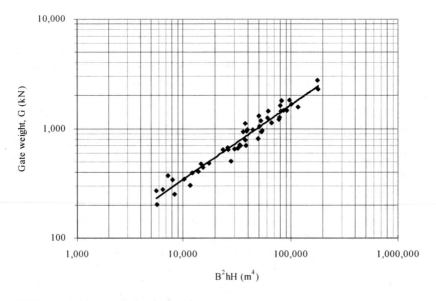

Fig. 7.1 Weight of spillway segment gates

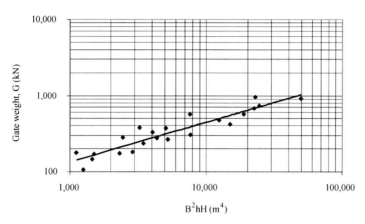

Fig. 7.2 Weight of submerged segment gates

7.3 FIXED-WHEEL GATES

The approximate weight of fixed-wheel gates is given by the expressions

$$G = 0.735 \, (B^2hH)^{0.697} \tag{7.3}$$

for $B^2hH > 2000$ m^4, and

$$G = 0.886 \, (B^2hH)^{0.654} \tag{7.4}$$

for $B^2hH < 2000$ m^4. These equations are shown together with a selection of actual weights in Figures 7.3 and 7.4, respectively.

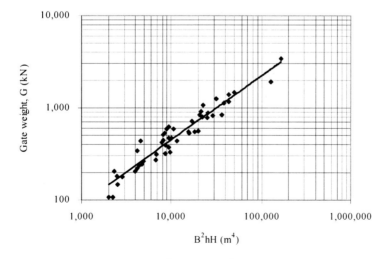

Fig. 7.3 Weight of large fixed-wheel gates ($B^2hH > 2000$m^4)

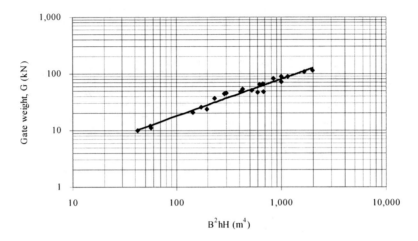

Fig. 7.4 Weight of small fixed-wheel gates ($B^2hH < 2000m^4$)

It was evident that, for values of B^2hH less than 12 000 m^4, fixed-wheel gates weigh less than segment gates. Equation 7.4, corresponding to small fixed-wheel gates ($B^2hH < 2000$ m^4), gives the best correlation to the analyzed data, when compared with the other types.

7.4 DOUBLE-LEAF FIXED-WHEEL GATES

The following expression gives the approximate weight of double-leaf fixed-wheel gates:

$$G = 0.913 \, (B^2hH)^{0.669} \qquad\qquad (7.5)$$

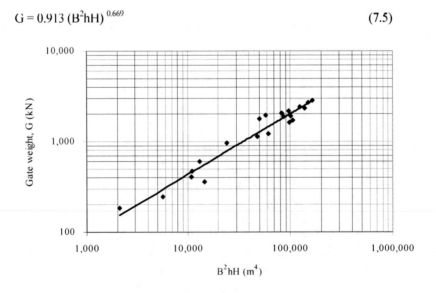

Fig. 7.5 Weight of double-leaf fixed-wheel gates

The expression is compared with actual gate weights in Figure 7.5. These gates, not used in submerged works, are approximately 25 per cent heavier than the segment-gate type. Notwithstanding this disadvantage, double-leaf fixed-wheel gates have been used successfully in Europe and Japan as they permit both the discharge over and under the gate leaf, as a result of lowering the top panel or lifting the lower panel, as desired. Both panels can also be lifted completely to allow the passage of the maximum discharge in the case of flood.

7.5 STOPLOGS

The weight of one complete set of stoplogs for one opening is given by the equation

$$G = 0.601 \, (B^2hH)^{\,0.703} \tag{7.6}$$

for spillway stoplogs, and

$$G = 0.667 \, (B^2hH)^{\,0.678} \tag{7.7}$$

for submerged stoplogs.

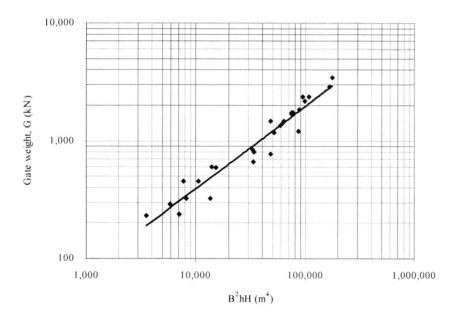

Fig. 7.6 Weight of spillway stoplogs

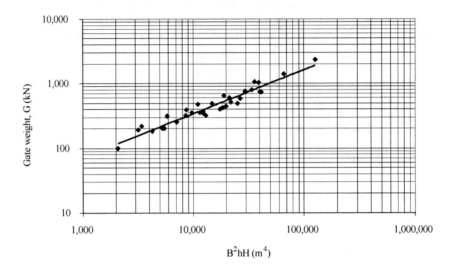

Fig. 7.7 Weight of submerged stoplogs

The weight of one stoplog panel is obtained by dividing the weight of the complete set by the number of panels, provided that all panels have the same height. Thus to determine the weight of one set of stoplogs, consisting of four panels 8 m wide, with a total height of 9 m, subjected to a head of 32 m on the sill:

$$B^2hH = 8^2 \cdot 9 \cdot 32 = 18432 \text{ m}^4$$

Hence, from Equation 7.7, the weight of one set of stoplogs is

$$G = 0.667 \cdot 18432^{0.678} = 520 \text{ kN}$$

So, each panel will weigh

$$G' = G/4 = 520/4 = 130 \text{ kN}$$

7.6 FLAP GATES

The weight of flap gates is estimated by the expression

$$G = 2.387 \, B(hH)^{0.643} \tag{7.8}$$

Graphically, the weight is obtained by Figure 7.8, where the abscissa represents the product *hH* (gate height x head on sill) and the ordinate shows the ratio *G/W* between the weight and the gate span. The following example shows how to obtain the weight of a flap gate 15 m wide and 3.5 m high, for a water head of 3.5 m. The product *hH* is 3.5 · 3.5 = 12.25 m². The graph gives *G/B* = 12 kN/m, giving a gate weight of *G* = 12 · 15 = 180 kN.

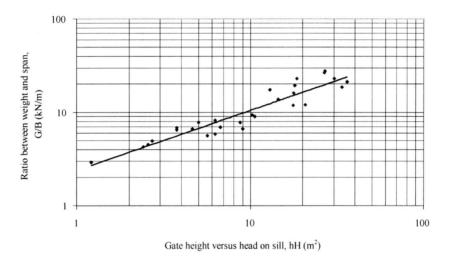

Fig. 7.8 Weight of flap gates

7.7 CATERPILLAR GATES

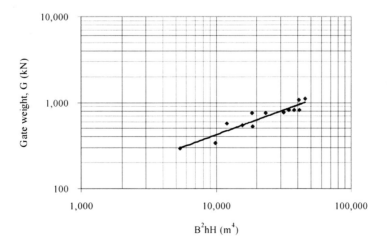

Fig. 7.9 Weight of caterpillar gates

The approximate weight of caterpillar gates is given by the expression

$$G = 2.102 \, (B^2hH)^{\,0.576} \qquad\qquad (7.9)$$

A selection of actual weights of caterpillar gates is shown in Table 7.10.

7.8 EMBEDDED PARTS

Data on the weight of embedded parts were also collected. Several attempts have been made to correlate the weight of the embedded parts to gate dimensions, length of guides and/or head on sill, but the results were unsatisfactory. A comparison between the weight of the embedded parts and the weight of the gate shows a wide variation. This results from the multiplicity of criteria adopted for the design of the embedded parts, such as the extension of the slot steel lining, length of main guides, allowable concrete bearing stress, and so on. In some cases, one set of embedded parts weighs as much as the gate, as in the case of high-head segment gates.

Table 7.1 summarizes the results of this study and indicates the range and the mean value of the ratio between the weight of the embedded parts and the weight of the gate. As can be seen, the embedded parts represent a considerable percentage of the total weight of supply, especially in the submerged gates. Although not conclusive, the mean values shown in Table 7.1 can be adopted to estimate the probable weight of the embedded parts.

Table 7.1 Weight of Embedded Parts

Gate type	Ratio between weight of embedded parts and weight of gate		
	Minimum	Maximum	Mean
Spillway segment	0.09	0.21	0.13
Submerged segment	0.15	1.10	0.60
Fixed-wheel with $B^2hH > 2000m^4$	0.10	0.50	0.32
Fixed-wheel with $B^2hH < 2000m^4$	0.11	0.54	0.30
Double-leaf fixed-wheel	0.10	0.40	0.21
Spillway stoplog	0.04	0.10	0.06
Submerged stoplog	0.07	0.36	0.18
Flap gate	0.07	0.78	0.30

Legend:
   B = span of gate, m
   h = height of gate, m
   H = head on sill, m.

Table 7.2 Spillway Segment Gates

	Year	Project	Span B (m)	Height h (m)	Head on sill H (m)	B^2hH (m^4)	Weight (kN)	Manufacturer
1	1982	Tucuruí	20.00	21.20	21.2	179776	2301	DEDINI
2	1982	Itaipu	20.00	21.34	20.84	177890	2756	DEDINI & others
3	1978	Salto Grande	15.30	22.33	22.33	116724	1579	ATB
4	1974	Salto Osório	15.30	20.77	20.77	100985	1667	Confab/Kurimoto
5	1979	Salto Santiago	15.30	20.77	20.00	97241	1815	Ishibrás
6	1977	Itaúba	15.00	20.30	20.00	91350	1471	Bardella
7	1986	Itaparica	15.00	19.70	19.40	85991	1472	Bardella
8	1974	Água Vermelha	15.00	19.60	18.60	82026	1805	ALSTOM
9	1977	São Simão	15.00	18.78	18.98	80200	1446	DEDINI
10	1980	Foz do Areia	14.50	19.45	19.45	79538	1620	DEDINI
11	1979	Itumbiara	15.00	18.86	18.36	77911	1275	DEDINI
12	1975	Marimbondo	15.00	18.85	18.35	77827	1216	DEDINI
13	1970	Jaguara	13.50	19.50	18.50	65747	1128	Krupp
14	1981	Colbún-Machicura	14.40	17.20	17.20	61345	1452	BYNSA
15	1996	Miranda	12.50	19.96	19.46	60691	1249	IMPSA
16	1972	Mascarenhas	13.80	16.87	16.87	54199	963	Ishibrás
17	1976	Capivara	15.00	15.60	15.20	53352	938	DEDINI
18	1969	Ilha Solteira	15.00	15.46	15.13	52630	1177	ALSTOM
19	1979	Paulo Afonso IV	11.50	19.65	19.65	51065	1049	DEDINI
20	1978	Emborcação	12.00	18.70	18.70	50355	1315	ALSTOM
21	1970	Porto Colômbia	15.00	14.90	14.90	49952	814	DEDINI
22	1978	San Lorenzo	12.20	17.30	17.30	44546	981	BYNSA
23	1971	Cedillo	14.00	14.16	14.16	39299	975	ATB
24	1968	Jupiá	15.00	13.70	12.70	39148	945	Ishibrás
25	1974	Sobradinho	13.75	13.40	15.14	38356	702	Voith
26		Balbina	13.50	15.30	13.60	37923	785	ALSTOM
27	1977	El Khattabi	10.50	18.50	18.50	37733	1108	BYNSA
28	1964	Funil	11.50	16.50	16.50	36005	942	Bardella
29	1964	Funil	13.00	14.22	14.22	34173	706	ALSTOM
30	1967	Estreito	11.50	16.20	15.80	33851	711	DEDINI
31	1961	Furnas	11.50	15.86	15.40	32301	661	I.H.I
32	1974	Coaracy Nunes	12.50	14.10	13.60	29963	650	DEDINI
33	1974	Volta Grande	15.00	11.41	10.90	27983	506	DEDINI
34	1968	Boa Esperança	13.00	12.50	12.50	26406	642	Ishibrás
35		Kafue Gorge	14.00	11.53	11.53	26056	677	Riva Calzoni
36	1960	Três Marias	11.00	14.00	14.00	23716	647	ALSTOM
37	1971	Passo Real	11.50	11.77	11.27	17543	483	Ishibrás
38	1977	Banabuiu	15.00	8.50	8.00	15300	441	Ishibrás
39		Zarga	10.00	12.10	12.10	14641	477	Riva Calzoni
40		El Chocon	14.00	8.41	8.410	13863	405	ATB
41		Tavera	9.20	12.00	12.00	12188	393	ATB
42	1968	Salto Mimoso	13.00	8.48	8.18	11723	304	Ishibrás
43	1972	Curuá-Una	10.00	10.26	10.00	10260	344	Ishibrás
44	1968	Salto Mimoso	11.00	8.48	8.18	8393	253	Ishibrás
45	1977	Eng. Avidos	10.00	9.00	9.00	8100	343	Ishibrás
46	1954	Cachoeira do Funil	10.00	8.50	8.62	7327	373	ALSTOM
47	1972	Paulo Afonso III	7.67	10.50	10.50	6486	277	Voith
48		Capivari-Cachoeira	10.00	8.00	7.14	5712	204	Bardella
49	1967	Foz do Chopim	12.50	6.00	6.00	5625	271	DEDINI

Table 7.3 Submerged Segment Gates

	Year	Project	Span B (m)	Height h (m)	Head on sill H (m)	B^2Hh (m⁴)	Weight G (kN)	Manufacturer
1	1979	Jebba	12.00	9.50	36.00	49248	908	Mitsubishi
2	1974	Sobradinho	9.80	7.50	33.87	24397	734	Voith
3	1962	Roseires	10.00	13.00	17.53	22789	954	Sorefame
4	1973	Moxotó	10.00	8.00	28.00	22400	675	Bardella/Sorefame
5	1958	Macagua	10.00	11.00	17.00	18700	571	VA TECH Hydro
6	1969	Dez	12.80	5.60	16.30	14955	419	Krupp
7		Aguieira	9.50	8.86	15.56	12442	476	Sorefame
8		Kariba	4.90	6.17	51.89	7687	305	Sorefame
9	1970	Midorikawa	6.20	6.30	31.40	7604	566	Mitsubishi
10		Picote	4.30	5.00	57.20	5288	265	Sorefame
11		Régua	5.00	5.80	35.15	5097	373	Sorefame
12	1969	Matsubara	4.40	4.40	51.60	4395	279	Hitachi-Zosen
13	1970	Lower Tachien	5.00	7.00	23.30	4078	333	Mitsubishi
14	1964	Tsuruta	4.30	4.15	45.60	3499	237	Hitachi-Zosen
15	1977	Oishi	3.63	5.43	45.63	3265	383	Hitachi-Zosen
16	1975	Palagnedra	5.10	4.30	26.00	2908	181	ZWAG
17		Valeira	4.20	3.80	36.57	2451	282	Sorefame
18	1971	Rio Prado	3.50	3.50	54.70	2345	176	VA TECH Hydro
19		Oguchigawa	4.62	3.40	20.80	1509	171	Hitachi-Zosen
20	1966	Tinajones	3.60	2.83	40.00	1467	147	M.A.N.
21	1969	Capivarí-Cachoeira	2.50	4.45	45.00	1252	107	Ishibrás
22		Yado	2.82	2.82	49.70	1115	178	Hitachi-Zosen

Table 7.4 Fixed-Wheel Gates with $B^2hH > 2\,000$ m⁴

	Year	Project	Span B (m)	Height h (m)	Head on sill H (m)	B^2hH (m⁴)	Weight G (kN)	Manufacturer
1	1982	Itaipu	7.31	22.37	137.00	163765	3433	ALSTOM
2	1979	Sobradinho	12.30	19.15	43.28	125391	1913	DEDINI
3	1975	Capivara	8.50	12.00	58.00	50286	1481	Bardella/Sorefame
4	1978	Água Vermelha	10.40	9.53	42.00	43292	1398	Bardella
5		Guri	6.40	15.00	70.00	43008	1177	BYNSA
6	1969	Ilha Solteira	8.50	9.00	58.50	38040	1128	ALSTOM
7	1979	Itumbiara	7.30	11.84	58.10	36658	834	Krupp
8	1964	Jupiá	11.65	7.50	31.00	31555	1256	Bardella & others
9	1981	Nova Avanhandava	6.54	18.00	37.80	29102	831	ALSTOM
10	1980	Foz do Areia	7.40	7.40	63.70	25813	883	Bardella
11	1976	Moxotó	7.10	14.30	34.90	25158	785	ALSTOM
12	1973	Porto Colômbia	10.85	5.43	35.60	22757	1079	ALSTOM
13		San Lorenzo	6.00	14.35	43.00	22214	795	BYNSA
14	1971	Volta Grande	6.33	14.80	36.00	21349	915	Bardella
15	1975	São Simão	6.50	11.30	43.64	20835	834	Voith
16	1979	Salto Santiago	6.60	11.10	41.60	20114	559	DEDINI
17	1968	Jaguara	7.20	9.70	36.30	18253	554	Coemsa/GIE
18	1980	Paulo Afonso IV	8.70	9.35	24.25	17162	716	Ishibrás
19	1955	Euclides da Cunha	6.50	6.60	57.25	15964	530	ALSTOM
20	1978	Emborcação	5.44	8.74	60.25	15584	549	ALSTOM
21		Malpaso	4.80	8.20	62.10	11732	438	ATB
22	1972	Promissão	9.10	6.20	20.80	10679	589	Voith
23	1968	Boa Esperança	6.75	6.68	33.00	10044	471	Ishibrás
24		Plover Cover	9.14	8.53	13.72	9777	329	Riva Calzoni

Table continues

Table 7.4 Fixed-Wheel Gates with $B^2hH > 2\,000$ m^4 (continued)

	Year	Project	Span B (m)	Height h (m)	Head on sill H (m)	B^2hH (m^4)	Weight G (kN)	Manufacturer
25		Bandama	5.70	10.00	29.20	9487	620	Riva Calzoni
26	1959	Três Marias	4.50	8.15	57.00	9407	471	I.H.I.
27		Colbún-Machicura	4.50	8.00	58.00	9396	373	BYNSA
28		Balbina	4.79	11.36	34.27	8932	589	ALSTOM
29	1970	Estreito	6.55	10.63	19.00	8665	319	Riva Calzoni
30	1973	Passo Real	5.10	9.00	37.00	8661	392	Bardella
31		La Angostura	3.45	8.76	81.00	8446	527	ATB
32	1973	Mascarenhas	4.56	12.17	32.67	8267	513	Ishibrás
33		La Villita	3.75	10.50	55.54	8201	444	ATB
34	1974	Salto Osório	7.40	7.40	19.70	7983	425	Ishibrás
35	1960	Furnas	4.60	9.65	33.50	6840	314	Krupp & others
36		Macchu Picchu	12.00	6.50	7.20	6739	272	ATB
37	1996	Miranda	5.50	7.65	21.46	4966	262	IMPSA
38	1966	Peixoto	6.00	6.00	22.00	4752	246	Ishibrás
39		Tachfine	3.50	5.25	71.00	4566	434	Riva Calzoni
40		El Novillo	3.60	6.60	51.75	4426	238	ATB
41		Colbún-Machicura	4.50	8.00	26.00	4212	343	BYNSA
42		Inga	5.50	5.50	25.10	4176	221	ATB
43		Santa Rosa	4.00	5.00	50.12	4010	205	ATB
44		Khasm-el-Girba	5.05	4.70	24.00	2877	180	ATB
45		Capivarí-Cachoeira	4.00	5.30	30.00	2544	146	Voith
46		Cedillo	3.40	4.50	48.00	2497	182	ATB
47		Peligre	2.84	6.35	45.30	2320	206	Riva Calzoni
48	1984	Mirorós	4.00	3.50	40.00	2240	108	Müller
49	1977	Itaúba	3.80	6.26	22.50	2034	108	Coemsa

Table 7.5 Fixed-Wheel Gates with $B^2hH < 2\,000$ m^4

	Year	Project	Span B (m)	Height h (m)	Head on sill H (m)	B^2hH (m^4)	Weight G (kN)	Manufacturer
1	1965	Guandu	8.00	5.30	5.80	1967	114	Ishibrás
2		Fadalto	3.25	4.80	32.22	1634	108	Riva Calzoni
3	1965	Guandu	8.00	4.00	4.50	1152	89	Ishibrás
4		Tucuruí	3.15	3.15	32.00	1000	72	DEDINI
5		La Spezia (Pos. 3-4)	4.40	7.40	6.90	989	89	Riva Calzoni
6	1978	Serraria	3.50	3.50	19.50	836	82	DEDINI
7	1996	Sá Carvalho	4.75	4.00	7.50	677	48	Usiminas
8	1965	Guandu	8.00	3.00	3.50	672	66	Ishibrás
9	1973	Alecrim	3.00	3.80	18.10	619	65	DEDINI
10	1961	Bariri	2.95	2.10	32.40	592	47	CKD-Blansko
11		Diga di Ganda	3.00	2.50	23.20	522	51	ATB
12		S. Fiorano	4.50	4.60	4.60	428	53	Riva Calzoni
13		Pisayambo	2.10	2.60	35.66	409	48	ATB
14		La Spezia (Pos. 1)	2.45	7.20	7.00	303	45	Riva-Calzoni
15	1968	Taipu	2.00	2.00	36.30	290	44	Ishibrás
16		Valdarda	1.70	1.70	47.35	233	37	Riva Calzoni
17	1996	Sá Carvalho	2.20	5.10	8.00	197	24	Usiminas
18		Ponte Liscione	2.50	2.50	11.00	172	26	Riva Calzoni
19		Soledade	1.68	2.60	19.50	143	21	Isomonte
20		Macchu Picchu	1.90	3.50	4.50	57	11	ATB
21	1965	Guandu	2.60	2.10	3.92	56	12	Ishibrás
22	1965	Guandu	2.70	2.15	2.65	42	10	Ishibrás

Table 7.6 Double-Leaf Fixed-Wheel Gates

	Year	Project	Span B (m)	Height h (m)	Head on sill h (m)	B^2hH (m⁴)	Weight G (kN)	Manufacturer
1	1965	Ybbs-Persenbeug	30.00	13.50	13.50	164025	2845	Waagner-Biro
2	1980	Crestuma	28.00	13.80	13.80	149305	2707	Sorefame
3	1961	Aschach	24.00	15.50	15.50	138384	2344	Waagner-Biro
4	1960	Beauchastel	26.00	13.50	13.50	123201	2403	ALSTOM
5	1963	Passau-Ingling	23.00	14.30	14.00	105906	1723	VA TECH Hydro
6	1967	Wallsee-Mitterkirchen	24.00	13.20	13.20	100362	1913	Waagner-Biro
7	1959	Scharding	23.00	13.80	13.50	98553	1634	VA TECH Hydro
8	1953	Braunau	23.00	13.50	13.50	96410	2178	Tissen-Klonne
9	1972	Ottensheim	24.00	12.50	12.00	86400	1876	VA TECH Hydro
10	1954	Jochenstein	24.00	12.10	11.80	82241	2050	VA TECH Hydro
11	1965	Komori	14.00	17.80	17.76	61961	1226	Mitsubishi
12		Belver	17.00	14.15	14.15	57864	1938	Sorefame
13	1951	Rosenau	16.00	13.50	14.50	50112	1783	VA TECH Hydro
14	1959	Losenstein	13.50	16.20	16.20	47830	1138	Waagner-Biro
15		Donzère-Mondragon	24.00	6.50	6.50	24336	958	
16	1968	Yamanoi	12.70	9.50	9.50	14556	358	Mitsubishi
17	1964	Miyagoochi	9.00	12.50	13.00	13163	600	Mitsubishi
18	1952	Kniepass	11.50	9.30	9.00	11069	469	VA TECH Hydro
19	1960	Ichibusa	7.10	14.60	14.80	10893	404	Mitsubishi
20	1966	Shimbashi	10.00	7.50	7.50	5625	248	Mitsubishi
21	1952	Rauris Kitzloch	8.00	6.00	5.50	2112	185	VA TECH Hydro

Table 7.7 Spillway Stoplogs

	Year	Project	Span B (m)	Height h (m)	Head on sill H (m)	B^2hH (m⁴)	Weight G (kN)	Manufacturer
1	1982	Itaipu	20.00	21.60	20.42	176429	3461	DEDINI & others
2	1984	Tucuruí	20.00	20.52	20.50	168264	2893	DEDINI
3		Salto Osório	16.00	21.00	20.00	107520	2354	Kurimoto
4	1979	Salto Santiago	15.30	21.00	20.00	98318	2188	Ishibrás
5	1977	Itauba	15.00	20.58	20.23	93675	2354	Vogg
6	1980	Foz do Areia	15.10	20.00	19.20	87556	1834	DEDINI
7	1985	Itaparica	15.00	20.10	19.10	86380	1202	DEDINI
8	1977	Itumbiara	15.00	18.75	18.26	77034	1748	DEDINI
9	1977	São Simão	15.00	18.45	18.00	74723	1727	DEDINI
10	1974	Marimbondo	15.00	18.30	18.00	74115	1682	Ishibrás
11		Jaguara	13.50	19.10	18.10	63006	1470	ALSTOM
12	1972	Porto Colômbia	15.00	15.25	18.00	61763	1403	Ishibrás
13	1996	Miranda	12.50	19.80	19.00	58781	1343	IMPSA
14	1975	Capivara	15.00	15.40	15.00	51975	1177	Bardella/Sorefame
15		Emborcação	12.00	18.48	18.00	47900	1472	Ishibrás
16		Pedra do Cavalo	15.30	14.57	14.00	47750	768	ALSTOM
17	1969	Estreito	11.50	16.00	16.00	33856	806	M.A.N.
18	1972	Promissão	11.20	16.50	16.20	33530	665	Bardella
19	1960	Furnas	11.50	15.75	15.75	32806	844	Krupp & others
20	1977	Banabuiu	15.00	8.50	8.00	15300	589	Ishibrás
21	1977	Eng. Avidos	10.00	12.00	11.62	13944	603	Ishibrás
22	1968	Boa Esperança	13.00	9.00	9.00	13689	324	Ishibrás
23	1972	Curuá-Una	10.00	10.50	10.17	10679	451	Ishibrás
24	1968	Salto Mimoso	13.00	7.00	7.00	8281	325	Ishibrás
25		Cachoeira do Funil	10.00	8.85	8.85	7832	451	ALSTOM
26	1961	Bariri	12.10	7.00	7.00	7174	238	CKD Blansko
27	1968	Salto Mimoso	11.00	7.00	7.00	5929	288	Ishibrás
28	1973	Moxotó	10.00	6.00	6.00	3600	230	ALSTOM

Table 7.8 Submerged Stoplogs

	Year	Project	Span B (m)	Height h (m)	Head on sill H (m)	B^2hH (m⁴)	Weight G (kN)	Manufacturer
1		Sobradinho	11.00	25.60	40.76	126258	2335	Bardella
2	1984	Tucuruí	10.10	14.45	45.00	66332	1409	DEDINI
3		Itaparica	9.50	12.15	38.00	41668	750	Ishibrás
4	1977	Itumbiara	7.30	13.14	57.69	40396	750	DEDINI
5	1982	Itaipu	10.00	9.90	40.00	39600	1040	Bardella & ALSTOM
6	1972	Promissão	10.00	12.00	30.30	36360	1057	Bardella
7	1974	Marimbondo	8.16	13.67	37.30	33951	811	Ishibrás
8	1972	Porto Colômbia	6.84	21.00	30.20	29671	755	Ishibrás
9	1976	Moxotó	11.95	8.96	23.16	29633	745	ALSTOM
10	1981	Nova Avanhandava	6.30	19.60	34.50	26838	589	ALSTOM
11	1979	Salto Santiago	7.00	12.35	42.00	25416	490	DEDINI
12	1977	São Simão	6.50	12.36	42.23	22053	516	DEDINI
13		Volta Grande	6.34	15.77	33.80	21425	605	Ishibrás
14	1972	Porto Colômbia	10.85	6.47	26.30	20032	444	Ishibrás
15	1980	Paulo Afonso IV	8.90	10.20	23.66	19116	649	Ishibrás
16	1980	Paulo Afonso IV	5.80	14.30	38.50	18521	423	ALSTOM
17	1969	Jaguara	7.26	10.15	32.80	17547	402	DEDINI
18	1981	Nova Avanhandava	10.10	7.71	18.79	14778	495	Bardella
19	1960	Furnas	5.66	12.14	33.50	13029	319	Krupp & others
20	1980	Foz do Areia	9.78	4.76	27.30	12429	369	Bardella
21	1977	São Simão	10.37	4.78	23.98	12326	349	DEDINI
22	1977	Itumbiara	8.77	5.90	25.46	11553	354	DEDINI
23	1974	Marimbondo	7.40	5.06	39.65	10986	477	Ishibrás
24	1968	Jupiá	7.56	8.24	20.77	9782	354	Ishibrás
25	1974	Cachoeira Dourada	6.80	9.25	20.30	8683	388	Ishibrás
26	1968	Boa Esperança	7.75	3.57	40.00	8577	320	Ishibrás
27	1973	Salto Osório	8.50	4.25	23.10	7093	257	Ishibrás
28	1969	Estreito	6.16	6.46	23.90	5859	314	Santa Matilde
29	1973	Mascarenhas	5.15	7.65	27.31	5541	201	Ishibrás
30	1996	Miranda	5.50	8.19	21.45	5314	204	IMPSA
31	1996	Miranda	7.00	4.20	20.79	4279	182	IMPSA
32	1974	Coaracy Nunes	5.55	6.54	17.00	3425	219	Ishibrás
33	1976	Curuá-Una	4.94	6.25	21.00	3203	193	Ishibrás
34	1977	Itauba	3.80	6.40	22.50	2079	98	Vogg

Table 7.9 Flap Gates

	Year	Project	Span B (m)	Height H (m)	Head on sill H (m)	Weight G (kN)	H h (m²)	Weight/Span G/B (kN/m)	Manufacturer
1		Promissão	8.00	6.00	6.00	169	36.00	21.13	Sorefame
2		Cabora Bassa	12.00	5.80	5.80	223	33.64	18.58	Sorefame
3	1958	Ottendorf	30.00	5.50	5.50	689	30.25	22.97	M.A.N.
4	1960	Lohmuhle	8.00	5.17	5.17	221	26.73	27.63	M.A.N.
5	1960	Altusried	10.00	5.15	5.15	265	26.52	26.50	M.A.N.
6	1959	Sihl-Hofe	8.50	4.55	4.55	102	20.70	12.00	ZWAG
7	1979	Riedenburg	15.00	4.30	4.30	343	18.49	22.87	M.A.N.
8		Biopio	15.00	4.25	4.25	288	18.06	19.20	Sorefame
9	1959	Innerferrera	8.00	3.70	4.80	129	17.76	16.13	ZWAG
10	1960	Thun	12.00	4.15	4.25	142	17.64	11.83	ZWAG
11	1974	Takase Zeki	6.00	3.80	3.80	82	14.44	13.67	Mitsubishi
12	1978	Kemnader See	25.00	3.60	3.60	436	12.96	17.44	Krupp & others
13	1962	Hausen	24.10	3.25	3.25	216	10.56	8.96	M.A.N.
14		Limpopo	13.30	3.20	3.20	125	10.24	9.40	Sorefame

Table continues

Table 7.9 Flap Gates (continued)

	Year	Project	Span B (m)	Height H (m)	Head on sill H (m)	Weight G (kN)	H h (m²)	Weight/Span G/B (kN/m)	Manufacturer
15	1955	Ottenstein	27.70	3.00	3.00	185	9.00	6.68	VA TECH Hydro
16		Matala	17.50	2.95	2.95	137	8.70	7.83	Sorefame
17	1962	Hale	6.10	2.59	2.59	42	6.71	6.89	VA TECH Hydro
18	1967	Muro-Matsubara	40.00	2.50	2.50	329	6.25	8.23	Mitsubishi
19	1968	Burfell	20.00	2.50	2.50	116	6.25	5.80	Krupp
20	1967	Gmunden	22.50	2.37	2.37	127	5.62	5.64	VA TECH Hydro
21		Toobetsu A	25.00	2.00	2.50	194	5.00	7.76	Hitachi-Zosen
22	1974	Nukui	6.00	2.00	2.30	40	4.60	6.67	Mitsubishi
23		Toobetsu B	25.00	1.70	2.20	161	3.74	6.44	Hitachi-Zosen
24		Kamigawara	14.00	1.70	2.20	95	3.74	6.79	Hitachi-Zosen
25	1970	Kuritsubo	27.00	1.50	1.80	132	2.70	4.89	Mitsubishi
26	1958	Altdorf	19.74	1.60	1.60	88	2.56	4.46	M.A.N.
27	1956	Greinsfurt	55.00	1.55	1.55	235	2.40	4.27	VA TECH Hydro
28	1978	Bou Hertma	8.60	1.10	1.10	25	1.21	2.91	M.A.N

Table 7.10 Caterpillar Gates

	Year	Project	Span B (m)	Height h (m)	Head on sill H (m)	B^2hH (m⁴)	Weight G (kN)	Manufacturer
1	1969	Cabora Bassa	8.90	10.96	52.19	45308	1106	Sorefame
2	1095	Guri	5.50	15.75	85.80	40878	818	Krupp
3	1969	El Chocon	10.00	10.07	40.36	40643	1071	Sorefame
4		Itaparica	9.50	11.00	38.00	37725	816	Bardella
5	1972	Valeira	8.75	14.06	32.00	34447	822	Sorefame
6	1972	Guri	6.00	9.70	90.00	31428	765	BYNSA
7	1977	Pocinho	8.00	12.16	30.00	23347	758	Sorefame
8	1972	Marimbondo	6.60	11.44	37.30	18588	527	Bardella/Sorefame
9	1965	San Luis	5.34	6.99	92.72	18481	756	Mitsubishi
10	1975	Aguieira	5.00	7.50	83.25	15609	540	Sorefame
11		Tsuruta	6.45	6.80	42.50	12023	573	Hitachi-Zosen
12		Funil	4.50	6.25	77.80	9847	338	ALSTOM
13	1965	Angat	3.15	8.00	68.00	5398	294	Mitsubishi

Table 7.11 Weight of Gates

Gate type	Gate weight	Coefficient of determination (R^2)
Spillway segment gate	$G = 0.640 (B^2hH)^{0.682}$	0.951
Submerged segment gate	$G = 3.688 (B^2hH)^{0.521}$	0.865
Fixed-wheel gate with $B^2hH > 2000m^4$	$G = 0.735 (B^2hH)^{0.697}$	0.921
Fixed-wheel gate with $B^2hH < 2000m^4$	$G = 0.886 (B^2hH)^{0.654}$	0.967
Double-leaf fixed-wheel gate	$G = 0.913 (B^2hH)^{0.669}$	0.949
Spillway stoplog	$G = 0.601 (B^2hH)^{0.703}$	0.937
Submerged stoplog	$G = 0.667 (B^2hH)^{0.678}$	0.928
Flap gate	$G = 2.387 B(hH)^{0.643}$	0.878
Caterpillar gate	$G = 2.102 (B^2hH)^{0.576}$	0.880

Legend:
G = weight of gate, kN
B = span of gate, m
h = height of gate, m
H = head on sill, m.

REFERENCES

1. Schreiber, G.P.: *Hydro Power Plants* (in Portuguese), Editora Edgard Blücher Ltda/Engevix, Rio de Janeiro (1977).
2. Davis, C.V. and Sorensen, K.E.: *Handbook of Applied Hydraulics*, McGraw-Hill, 3rd edition (1969).
3. Boissonnault, F.L.: Estimating Data for Reservoir Gates, *Transactions of ASCE*, Paper No. 2352 (1948).
4. Gomes Navarro, J.L. and Juan Aracil, J.: *Saltos de Agua y Presas de Embalse*, Madrid, Spain (1964).
5. Erbisti, P.C.F. and Liu, M.: Heightening and Structural Reinforcement of Furnas' Spillway Segment Gates (in Portuguese), *Construção Pesada*, No. 87 (1978).

Author's note: This chapter was first published by the author in the International Water Power and Dam magazine on May 1984, under the title "Estimating Gate Weights". It is reproduced with the Editor's agreement.

Chapter 8

Hydrodynamic Forces

8.1 INTRODUCTION

When a gate is totally closed and the water is at rest, the pressures obey the hydrostatic laws and the hydraulic forces are easily determined by analytical methods. In the absence of any flow, the calculation of the vertical component of the hydraulic forces on the gate comprises solely the determination of its buoyancy. This static condition is characterized by a uniform value of the piezometric head. When the gate is partly open, the hydrostatic balance is broken and a non-uniform distribution of the piezometric head in the conduit, near the gate, is observed. The high flow velocities at the bottom surface of the gate, which reduces the local pressure, cause this phenomenon.

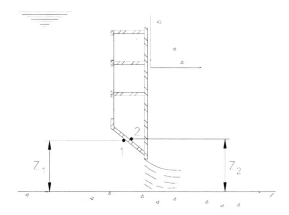

Fig. 8.1 Vertical lift gate partly open

The vertical lift gate of the Figure 8.1 is shown partly open. Applying the Bernoulli equation to points 1 and 2, we get:

$$\frac{P_1}{\gamma} + Z_1 + \frac{V_1^{\,2}}{2g} = \frac{P_2}{\gamma} + Z_2 + \frac{V_2^{\,2}}{2g}$$ (8.1)

Assuming that the water is at rest at the point 2, $V_2 = 0$.
As Z_1 is practically equal to Z_2, Equation 8.1 becomes

$$\frac{P_1}{\gamma} + \frac{V_1^{\,2}}{2g} = \frac{P_2}{\gamma}$$ (8.2)

or

$$\gamma \frac{V_1^{\,2}}{2g} = P_2 - P_1$$ (8.3)

As the left side of the equation is positive, there is a pressure difference between points 1 and 2, which increases with the velocity of the water flowing under the gate. The pressure at point 2 is practically equal to the hydrostatic pressure. Therefore, the gate bottom is subjected to a pressure difference, which causes a vertical downward force, called the downpull force. However, these hydrodynamic forces are not present in vertical lift gates with an upstream skin plate or in segment gates. F. H. Knapp [1] states the following rule: all surfaces of a gate located in regions of high water velocity and which form a sharp slope with the direction of the corresponding motion present the possibility of formation of hydrodynamic forces. In a segment gate, for example, at every point of the wet surface of the skin plate, the water flow has the same direction of movement of the gate, as shown in Figure 8.2.

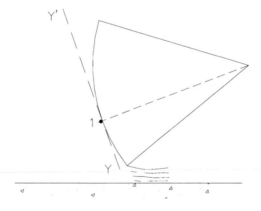

Fig. 8.2 Segment gate partly open

Since the water velocity has the same direction of movement as the gate at point 1 (tangent Y-Y'), there is no formation of hydrodynamic forces.

Figure 8.3 shows a vertical lift gate with an upstream skin plate.

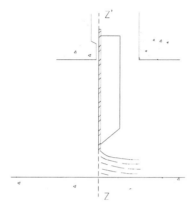

Fig. 8.3 Vertical lift gate with an upstream skin plate

The skin plate moves in the direction Z-Z', which is the same as the flow direction. Therefore, according to Knapp's rule, there is no formation of low-pressure areas and, consequently, thrust forces are not created.

The same does not apply to a vertical lift gate with a downstream skin plate (Fig. 8.4).

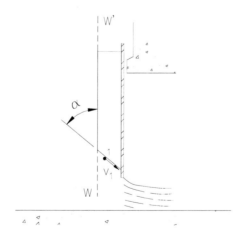

Fig. 8.4 Vertical lift gate with downstream skin plate

Any point of the gate bottom moves parallel to the direction *W-W'*, while the flow makes an angle α with *W-W'*. According to Knapp, this is the basic condition for creation of hydrodynamic forces.

8.2 MODEL TESTS

Vertical lift gates are frequently used for emergency closure in large penstocks for they can be designed to close under gravity, without the aid of hoists. This is very practical and safe because it allows the interruption of the water flow even in the event of energy failure. When the gates are required to operate in such a way, a careful appraisal of the downpull or uplift forces is needed, as such forces may cause serious operational problems, including catapulting or non-closure of the gate.

The only reliable method for determining the hydrodynamic forces is the gate model test. Such tests are made in hydraulic laboratories and may be conducted according to three different methods:

a) Direct measurement of forces through a force parallelogram, such as the one used in the test of intake gates of the Estreito Power Plant (Furnas), made by the hydraulic laboratory of Riva Calzoni, in 1968;

b) Direct measurement through a force transducer inserted in the suspension system of the gate model. Bardella used this equipment in the test of the Itumbiara Power Plant diversion gates, in 1977.

c) Pressure distribution method - pressures are measured by means of water manometers connected to piezometers in the top and bottom surfaces of the gate model. The product of the average pressure by the area gives the force acting normal to the surface. The difference between the vertical components of these forces gives the net downpull. In this method, the gate model is generally supported by a threaded stem, which keeps it in the desired open position. However, it is not possible to conduct dynamic tests with the gate. The Hydraulic Laboratory of the São Paulo University made this test in 1970 for the intake gates for the Passo Real Power Plant (CEEE), manufactured by Bardella.

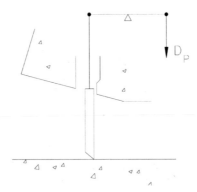

Fig. 8.5 Direct measurement of vertical downpull forces

Fig. 8.6 Test facilities for model test of the Passo Real Power Plant gates

The gate model tests are carried out with geometric similarity between the prototype and the model, with scales usually in the range of 1:20 to 1:30. The conversion of the geometric and hydraulic values from the model to the prototype is done according to Froude's similitude law, with the following factors:

. lengths, displacements k
. velocity, time $k^{1/2}$
. flow rate $k^{5/2}$
. forces k^3

where k is the scale relation between the prototype and the model.

The equipment and accessories of a typical test facility for hydraulic model tests comprise:

a) constant level reservoir, composed of supply pipe, tranquilizing chambers, weir, inlet and outlet valves;

b) gate model;

c) gate model drive mechanism, comprising a micro-motor, steel wire, sheaves, a force transducer and a gate displacement meter. Both transducer and displacement meter, when required, emit variations of the electric current that, after amplified, are recorded on paper by a Cartesian register. Thereby, the graphic force versus displacement is directly obtained;

d) return canal;

e) additional instrumentation, such as flow rate and reservoir level meters. These may be point gages, with a luminous electronic alarm to automatically show the contact between the tip and the water head.

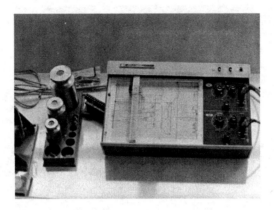

Fig. 8.7 Gate model position plotter

Local hydraulic phenomena in the slots and gate bottom, as well as in the reservoir (vortices etc), may be observed through transparent walls built of acrylic. The characterization of such phenomena is usually carried out by means of photographs.

Fig. 8.8 Model test of the Cabora Bassa dam gates (SOGREAH)

8.3 FACTORS INFLUENCING DOWNPULL

Experimental data of various hydraulic model investigations prove that the downpull is a function of the total head upstream of the gate and of the gate cross-sectional area, varies with the gate opening and undergoes the influence of a series of factors, such as:
. gate bottom geometry - sloping angle, radius of curvature of the upstream bottom edge, length of lip extension beneath the bottom horizontal beam etc;
. downstream projection of the top seal assembly:
. clearances between the gate and the shaft at partial openings;
. gate thickness;
. recess in the downstream wall of the gate shaft.

A detailed description of the influence of these factors is found in references [2, 3, 4]. There follows a summary of the influence of the main factors.

- Gate bottom geometry - for flat bottom gates (Fig. 8.9) the experimental data show that an increase in the e/d ratio reduces downpull, gate vibration and air demand. The reason for the downpull decrease is the reduction of the velocity near the bottom beam, which increases the local pressure. However, long lip extension requires considerable structural strengthening of the skin plate below the bottom beam. In practice, the e/d ratio is kept below 0.6.

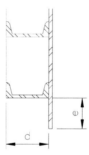

Fig. 8.9 Flat bottom gate

Fig. 8.10 Rounded bottom gate

In gates with curvature on the upstream bottom edge, an increase in the r/d ratio increases the pressures on the gate bottom, which consequently reduces the downpull. Large r/d ratios tend to cause flow instabilities and vibrations in the gate. As a rule, r/d not exceeding 0.5 is adopted as a practical limit. For gates with a flat sloping bottom (see Fig. 8.11), the experimental data shows that the magnitude of the downpull decreases as the θ - angle is increased. Structural and fabrication requirements do not favor adoption of $\theta >$ 45°.

Fig. 8.11 Sloping bottom gate

- Projection of the top seal - the magnitude of the downpull increases as d'/d is increased, where d' is the width of the skin plate and the top seal assembly and d is the gate thickness (Fig. 8.12). For intermediate gate positions, the area Bd (where B is the gate width) is subjected to nearly atmospheric air pressure at the bottom and to water pressure above. This pressure differential is responsible for the creation of a substantial part of the total downpull. In some designs of emergency gates, sub-atmospheric pressures are allowed downstream of the gate under free discharge conditions, in order to reduce the dimensions of the air vents. Thereby, the downpull in the Bd' area will be still greater. For gravity-closure gates the increase of the d'/d ratio may prove to be an useful tool in producing adequate downpull to overcome the friction forces which are maximum at small gate openings.

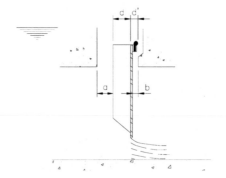

Fig. 8.12 Top seal projection and clearances between gate leaf and gate shaft

- Clearance between gate and shaft - experimental data demonstrate that downpull decreases as *a/b* is increased. The influence of *a/b* on the water levels in the gate shaft is shown in Figure 8.13. It will be observed that the greater is *a/b*, the higher the water depth in the shaft.

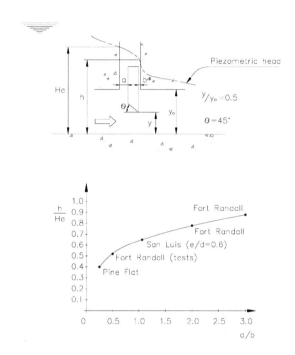

Fig. 8.13 Typical variation of h/H_e for various a/b

- Recess in the downstream wall of the gate shaft - a recess provided in the vertical wall downstream of the gate is effective in reducing downpull, since it tends to balance the water pressures above and below the top seal. Experimental studies showed that the recess performs its desired function if its depth is at least three times greater than the seal projection [4].

Fig. 8.14 Recess in downstream wall of gate shaft

8.4 FORMULAE FOR THE PREDICTION OF DOWNPULL

Gate downpull is normally determined using an equation such as:

$$D_p = \gamma \, K \, A \, H \qquad \text{(in kN)} \qquad\qquad (8.4)$$

where
γ = specific weight of water, in kN/m^3
K = downpull coefficient
A = cross-sectional area of gate, in m^2
H = operating head on the gate bottom, in meters.

The results of gate model tests permit one to establish the diagram K versus gate opening, for a given installation. Two papers published in the USA [2, 5] suggest K-empirical curves vs. gate opening. Tests made by the U. S. Corps of Engineers were based on 45 degrees sloping gate bottom, while the BUREC consider flat bottom gates with lip extension bellow the lower horizontal beam. However, both papers can only be used in the determination of the downpull of geometrically similar installations, which is rare in practice.

Figure 8.15 shows six charts of the K downpull coefficient variation with gate opening, based on results of model tests actually conducted with fixed-wheel gates with downstream seals and downstream skin plate. (The Marimbondo gates are of the caterpillar type). The gate bottom shape is shown in the figure, next to the corresponding chart. The main features of these gates are in Table 8.1.

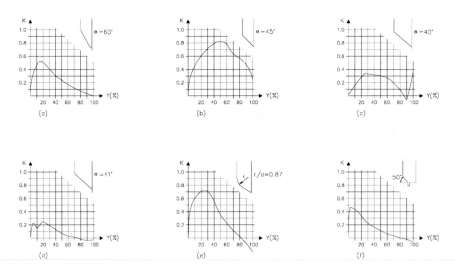

Fig. 8.15 Typical variation of the downpull coefficient K with gate opening Y for various projects. (a)Furnas; (b) Estreito; (c) Itumbiara diversion gate; (d) Itumbiara Intake gate; (e) São Simão; (f) Marimbondo

All tests were conducted with constant headwater corresponding to the maximum normal water level. The Furnas and Estreito tests were made under static conditions, i.e., with measurements made with the gate stopped in various positions, through readings in dynamometers. In the other tests (Itumbiara, São Simão and Marimbondo), transducers installed in the gate model suspension device determined the vertical forces and the measurement of forces was made with the gate in motion. An analysis of the charts shows that in four cases uplift forces occurred, not very intense, in the range of 80 to 100 per cent opening, a phenomenon characterized by negative values of K. Maximum downpull took place in the range of 15 to 50 per cent opening and the maximum K values for each gate varied from 0.25 (Itumbiara intake gates) to 0.83 (Estreito).

Table 8.1 Gate Model Tests

Project	Function	Supplier	Gate dimensions (m)		Head (m)	Model scale	Area (m^2) (*)	D_p/P_c (**)
			Span	Height				
Furnas	Intake	M.A.N.	4.60	9.65	33.5	1:20	3.49	1.8
Estreito	Intake	COEMSA	6.55	10.63	19.5	1:25	4.93	1.82
Itumbiara	Diversion	Bardella	5.00	7.86	25.75	1:20	6.3	1.24
Itumbiara	Intake	Krupp	7.30	11.84	58.1	1:30	10.57	1.76
São Simão	Intake	Voith	6.50	11.30	43.8	1:25	6.48	2.39
Marimbondo	Intake	SOREFAME	6.60	11.44	38.36	1:30	7.35	2.55

(*)   area of the horizontal projection of the gate
(**)   ratio between the maximum downpull and the gate weight

Such divergent results serve to emphasize the need of conducting specific model tests for every new installation, in order to predict the downpull forces acting on the gate. In the early stages of design such tests have not yet been conducted, but the determination of the hydraulic forces is necessary all the same. Therefore, in this phase, the use of test results of similar gates or resort to surveys is allowed, with the recommendations of adopting conservative values.

Naudascher, Kobus and Rao set forth, in their article *Hydrodynamic Analysis for High-Head Leaf Gates* [6], an analytical method for determination of the downpull forces based on geometric parameters of the gate and on the velocity in the contracted jet under the gate. According to the authors, the main portion of the downpull results from the difference of the integrated distributions of the piezometric head along the top and the bottom surfaces of the gate and is determined by:

$$P_1 = \left(K_T - K_B\right) B \ d \ \gamma \ \frac{V_j^2}{2g} \tag{8.5}$$

where:

K_T, K_B = top and bottom downpull coefficients
B = width of the gate (see Figure 8.16)
d = gate thickness (see Figure 8.16)
γ = specific weight of water
V_j = velocity in the contracted jet under the gate

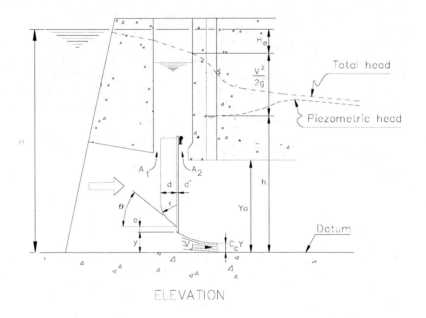

ELEVATION

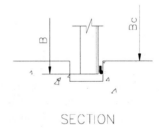

SECTION

Fig. 8.16 Geometric parameters for tunnel gates

Coefficients K_T and K_B are given by:

$$K_T = \frac{1}{B\,d} \int_0^d \int_0^B \frac{h_T - h}{\frac{V_j^2}{2g}}\, dB\, dx = \frac{1}{1 + \left(\dfrac{C_2\,A_2}{C_1\,A_1}\right)^2} \qquad (8.6)$$

and

$$K_B = \frac{1}{B\,d} \int_0^d \int_0^B \frac{h_j - h}{\frac{V_j 2}{2g}}\, dB\, dx \qquad (8.7)$$

where:

h_T = piezometric head on the top surface of the gate

h_i = piezometric head at a point on the gate bottom

h = piezometric head in the contracted jet

A_1 = cross-sectional area between upstream face of the gate and upstream wall of the gate shaft

A_2 = cross-sectional area of the contracted jet issuing from the gap between the downstream face of the gate and the downstream wall of the gate shaft

C_1, C_2 = discharge coefficients pertinent to the flow over the top of the gate (areas A_1 and A_2, respectively)

Given C_1, C_2, A_1 and A_2, the K_T coefficient may be graphically determined with the help of the Figure 8.17.

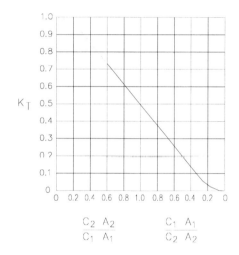

Fig. 8.17 Effect of the flow around the top of a tunnel-type gate on coefficient K_T

For face-type gates, the area $A_1 \rightarrow \infty$ and K_T is equal to unity. Thus, Equation 8.5 can be rewritten as

$$P_1 = (1 - K_B) B \, d \, \gamma \, \frac{V_j^2}{2g} \tag{8.5a}$$

Experiences carried out by the quoted authors revealed a good correspondence between the analytical method proposed and the results obtained in the model tests. Some of these results are shown in Figures 8.18 through 8.21 where the influence of various geometric parameters in the coefficient K_B may be noted.

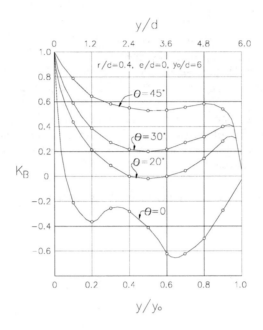

Fig. 8.18 Variation of K_B with relative gate opening for various θ

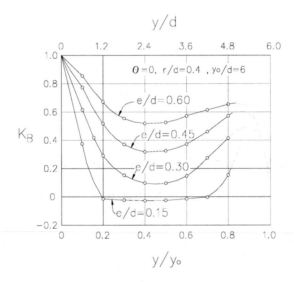

Fig. 8.19 Variation of K_B with relative gate opening for various e/d

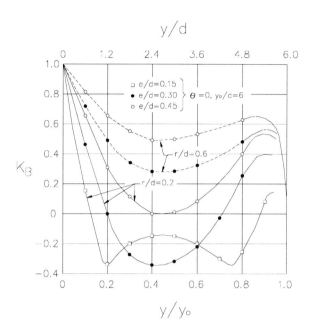

Fig. 8.20 Variation of K_B with relative gate opening for various r/d

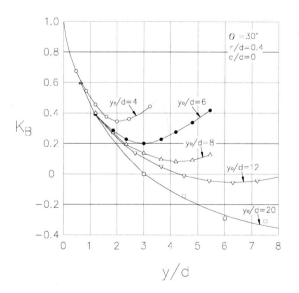

Fig. 8.21 Variation of K_B with relative gate opening for various Y_o/d

Data for K_B were obtained by graphical integration of the measured distribution of the piezometric head along the bottom surface of the gate according to Equation 8.7. The magnitude of the downpull force may be determined using K_B, for the design of the gate lip, and using K_T, for dimensioning of the flow passages around the gate top. The latter control is very important in tunnel-type gates, since slight changes in the ratio of the areas A_1 and A_2 affect substantially the primary portion P_1 of the downpull.

A second part of the downpull arises from the difference in pressure acting in horizontal projections of the gate as the top seal. If there is no flow between the gate and the lintel, that is, if the top seal is pressed against the downstream wall, A_2 is equal to zero and the downpull on the top seal is given by

$$P_2 = K_T \; A_s \; \gamma \; \frac{V_j^2}{2 \; g} \tag{8.8}$$

where A_s is the area of the horizontal projection of the top seal. On the other hand, if a recess is provided in the vertical wall downstream of the gate, $A_2 \neq 0$, and the water flows around the top seal and tends to equalize the pressures above and below the top seal, thus this force is eliminated.

For gates with an extended skin plate as shown in Figure 8.16, a third part of the downpull arises. It is calculated by the formula

$$P_3 = K_T \; d' \; \gamma \; \frac{V_j^2}{2 \; g} \tag{8.9}$$

where d' is the skin plate thickness.

The velocity in the vena contracta is given by

$$V_j = \sqrt{2 \; g \; (H - H_e - h)} \tag{8.10}$$

where
    H is the total head in the reservoir
    H_e embodies the entrance head-loss
    h is the piezometric head at the top of the vena contracta.

For free-surface flow, the piezometric head in the vena contracta is given by

$$h = C_c \, y + H_d \tag{8.11}$$

where
    C_c is the coefficient of contraction of the vena contracta
    y is the gate opening
    H_d is the depression downstream of the gate.

The velocity in the contracted jet may then be determined as a function of the gate opening through the equation

$$V_j = \sqrt{2g(H - H_e - C_c y - H_d)} \tag{8.12}$$

Its maximum value occurs at the start of opening ($y \approx 0$).

For submerged flow, the influence of the downstream water head should also be considered when determining the piezometric head in the vena contracta.

If the installation is provided with a control section at the downstream end of the gate (a turbine, for example), the gate flow increases according to the gate opening until peak controlled flow is attained. Starting from that point, the jet velocity is determined as a function of the opening y and of the maximum flow Q (assuming, to simplify, that flow exists only underneath the gate), using the equation

$$V_j = \frac{Q}{C_c \, A_j} = \frac{Q}{C_c \, B \; y} \tag{8.13}$$

where A_j is the cross-sectional area of the vena contracta.

Equalizing expressions 8.12 and 8.13, one gets, by trial and error, the opening y corresponding to the maximum control discharge.

The vena contracta coefficient (or gate discharge coefficient) C_c depends on the gate opening and geometry; it can be determined, in each case, through measurements in gate model tests. In the absence of specific values, the following coefficients may be adopted (whatever the bottom gate shape), suggested by the U. S. Army Corps of Engineers [7].

Table 8.2 Discharge Coefficients for Vertical Lift Gates

% of gate opening	10	20	30	40	50	60	70	80	90
Discharge coefficient, C_c	0.73	0.73	0.74	0.74	0.75	0.77	0.78	0.80	0.80

Example 8.1 Calculate the downpull on the following gate, knowing that the maximum headwater over the sill is 25.75 m and that the maximum discharge to control is 150 m³/s. Disregard the entrance head losses and the depression downstream of the gate.

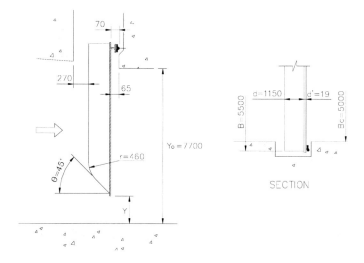

Fig. 8.22

Solution:

- Coefficient K_T:

The analysis of the Equation 8.6 shows that the relation between areas A_2 and A_1 exerts on the coefficient K_T a much greater influence than C_2/C_1 because, while A_2 and A_1 may take on any values (depending solely on the designer), practice demonstrates that the variation range of the discharge coefficients is small.
Consider the discharge coefficients C_1 and C_2 as equal.

Hence, Equation 8.6 gives:

$$K_T = \frac{1}{1 + \left(\dfrac{5500 \cdot 65}{5500 \cdot 270}\right)^2} = 0.95$$

- Coefficient K_B:

$r/d = 460/1150 = 0.4$
$e/d = 0/1150 = 0$
$Y_o/d = 7700/1150 = 6.7$

Figure 8.18 gives, for $\theta = 45$ degrees, $r/d = 0.4$ and $Y_o/d = 6$, the following values for K_B:

Y/Y_o	0.1	0.2	0.3	0.4	0.5	0.6	0.7	0.8	0.9
K_b	0.79	0.65	0.58	0.55	0.53	0.55	0.56	0.58	0.54

-    Velocity V_j:

At the start of opening, the velocity is maximum and equal to

$$V_j = \sqrt{2 \cdot 9.81(25.75 - 0 - 0 - 0)} = 22.48 \text{ m/s.}$$

Continuing the gate opening, the velocity becomes (Eq. 8.12):

$$V_j = \sqrt{2 \cdot 9.81(25.75 - C_c\ y)} = 4.429\sqrt{25.75 - C_c\ y}$$

At the maximum discharge, the velocity will be, according to Equation 8.13,

$$V_j = \frac{150}{5.5\,C_c\ y} = \frac{27.273}{C_c\ y}$$

Equalizing both expressions,

$$4.429\sqrt{25.75 - C_c\ y} = \frac{27.273}{C_c\ y}$$

By trial and error, $C_c y = 1.245$m. For an average discharge coefficient $C_c = 0.76$,

y = 1.245/0.76 = 1.638 m

This opening corresponds to $Y/Y_o = 1.638/7.7 = 0.213$, i.e., about 20 per cent of the maximum opening. From that point on, the jet velocity is calculated through Equation 8.13. Resorting to Table 8.2, the following table can be set:

Y/Y_o	-	0.1	0.2	0.3	0.4	0.5	0.6	0.7	0.8	0.9
y	m	0.77	1.54	2.31	3.08	3.85	4.62	5.39	6.16	6.93
V_j	m/s	22.48	22.48	13.03	9.77	7.71	6.26	5.30	4.52	4.02

- Force P_1:

Thus

$K_T = 0.95$
$B = 5.5$ m
$\gamma = 1000$ kgf/m^3 = 9810 N/m^3 = 9.81 kN/m^3
$g = 9.81$ m/s^2
$d = 1.15$ m

Equation 8.5 gives, in kN:

$$P_1 = (0.95 - K_R)\, 5.5 \cdot 1.15 \cdot 9.81\, \frac{V_j^{\,2}}{2 \cdot 9.81} = 3.16\,(0.95 - K_R)\, V_j^{\,2}$$

- Force P_2:

The projected area of the top seal assembly is

$A_s = 5.5 \cdot 0.07 = 0.39$ m^2.

According to Equation 8.8,

$$P_2 = 0.95 \cdot 0.39 \cdot 9.81\, \frac{V_j^{\,2}}{2 \cdot 9.81} = 0.185\ V_j^{\,2}$$

Due to the recess on the downstream vertical wall of the gate shaft (see Figure 8.22), force P_2 acts only as long as the upper seal remains pressed against the lintel. By way of simplification, suppose that P_2 acts only from the start of the opening ($y = 0$) to $y = 0.77$ m, i.e. that the upper seal remains pressed against the lintel in the first 77 cm of opening (10 per cent opening). Whereby,

$P_2 = 0.185 \cdot 22.48^2 = 93.49$ kN.

- Force P_3:

Equation 8.9 gives

$$P_3 = 0.95 \cdot 5.5 \cdot 0.019 \cdot 9.81 \frac{V_j^2}{2 \cdot 9.81}$$

Substituting in the preceding expressions the values of V_j and K_B, gives:

Y/Y_o	-	0.10	0.20	0.30	0.40	0.50	0.60	0.70	0.80	0.90
K_B	-	0.79	0.65	0.58	0.55	0.53	0.55	0.56	0.58	0.54
V_j	m/s	22.48	22.48	13.03	9.77	7.71	6.26	5.30	4.52	4.02
P_1	kN	335.35	558.92	225.33	135.73	88.29	55.72	39.06	27.12	23.49
P_2	kN	93.49	-	-	-	-	-	-	-	-
P_3	kN	25.27	25.27	8.49	4.77	2.97	1.96	1.40	1.02	0.81
$F_h =$ $P_1+P_2+P_3$	kN	454.11	584.19	233.82	140.50	91.26	57.68	40.46	28.14	24.30

Figure 8.23 shows the variation of the downpull forces F_h as a function of the gate opening.

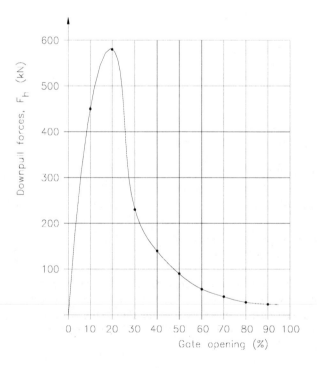

Fig. 8.23 Downpull forces – Example 8.1

8.5   METHOD OF KNAPP

F. H. Knapp developed a very simple method for analytical prediction of the hydrodynamic forces, which gives fairly conservative results. The gate is treated as having a constant opening along its thickness, and an 'equivalent' gate leaf thickness concept is introduced [1]. This varies along the gate travel opening and is determined in function of the bottom gate shape and the sill geometry. It is remarkable that Knapp analyzes only face-type gates and does not make any mention of the gaps between the gate and the shaft walls. The influence of the projection of the top seal and the skin plate is not considered.

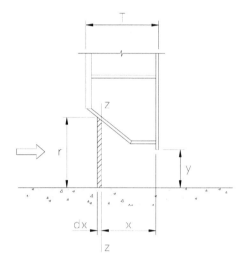

Fig. 8.24

The velocity of water at section Z-Z is

$$V = \frac{Q}{B\,r}$$

where
   Q = discharge rate
   r = gate opening at point x
   B = gate width.

and the velocity head is

$$\frac{V^2}{2\,g} = \left(\frac{Q}{B\,r}\right)^2 \frac{1}{2\,g}$$

Substituting $(Q/B)^2/2g$ by a constant C, one obtains

$$\frac{V^2}{2g} = \frac{C}{r^2} \tag{8.14}$$

According to Knapp, the local hydrodynamic force is equal to the kinetic energy at that point. So, the total hydrodynamic force acting downward on the bottom surface of the gate will be calculated by

$$P = \gamma B \int_0^T \frac{V^2}{2g} dx = \gamma \, BC \int_0^T \frac{dx}{r^2} \tag{8.15}$$

where T is the total thickness of the gate leaf.

At the minimum opening section $(r = y)$,

$$P = \gamma B \int_0^T \frac{dx}{y^2} = \gamma BC \frac{T_e}{y^2} \tag{8.16}$$

where T_e is the *'equivalent thickness'* of the gate leaf. T_e varies in function of the gate opening y.

The equivalent thickness can be determined by comparing Equations 8.15 and 8.16:

$$T_e = y^2 \int_0^T \frac{dx}{r^2} \tag{8.17}$$

At the minimum opening section, Eq. 8.14 gives

$$\frac{Vj^2}{2g} = \frac{C}{r^2}$$

that is,

$$C = \frac{Vj^2}{2g} y^2$$

Equation 8.16, becomes, then

$$P = \gamma B \left(\frac{Vj^2}{2g}\right) y^2 \frac{T_e}{y^2} = \gamma B \left(\frac{Vj^2}{2g}\right) T_e \tag{8.18}$$

The downward hydrodynamic forces can be calculated from Equations 8.17 and 8.18.

- Equivalent thickness

Case 1 Sloped sill

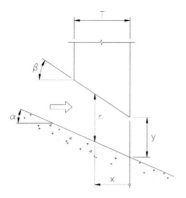

Fig. 8.25

The figure gives

r = y - x tanα + x tan β = y + x (tanβ - tanα)

By substituting *r* in Equation 8.17, the equivalent thickness will be

$$T_e = y^2 \int_0^T \frac{dx}{\left[y + x\left(\tan\beta - \tan\alpha\right)\right]^2} = \frac{T\ y}{T\ \left(\tan\beta - \tan\alpha\right) + y} \qquad (8.17a)$$

Case 2 Horizontal sill

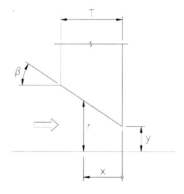

Fig. 8.26

In this case,

$$r = y + x \tan \beta$$

Using this value in Equation 8.17, it follows that

$$T_e = y^2 \int_0^T \frac{dx}{(y + x \tan\beta)^2} = \frac{T \; y}{T \; \tan\beta + y} \tag{8.17b}$$

Example 8.2 Using the Knapp method, determine the downward hydrodynamic forces acting on the gate of Example 8.1.

Solution:

Given that

$$T = d + d' = 1.150 + 0.019 = 1.169 \text{ m}$$
$$\beta = \theta = 45°$$

Since the sill is horizontal, the equivalent thickness can be calculated as per Equation 8.17b:

$$T_e = \frac{1.169 \; y}{1.169 \tan 45° + y} = \frac{1.169 \; y}{1.169 + y}$$

Therefore, the values of the equivalent thickness are determined as a function of the gate travel opening (y) as follows:

y	m	0.77	1.54	2.31	3.08	3.85	4.62	5.39	6.16	6.93
T_e	m	0.464	0.665	0.776	0.847	0.897	0.933	0.961	0.983	1.00

According to Equation 8.18, the hydrodynamic force will be calculated by

$$P = 9.81 \cdot 5.5 \frac{V_j^2}{2 \cdot 9.81} T_e = 2.75 \; V_j^2 \; T_e$$

Using the values of V_j calculated for the Example 8.1, one obtains

y	m	0.77	1.54	2.31	3.08	3.85	4.62	5.39	6.16	6.93
T_e	m	0.464	0.665	0.776	0.847	0.897	0.933	0.961	0.983	1.00
V_j	m/s	22.48	22.48	13.03	9.77	7.71	6.26	5.30	4.52	4.02
P	kN	644.80	924.20	362.30	222.30	146.60	100.50	74.20	55.20	44.40

By comparing the above results with those given in Example 8.1 (P versus P_l), it can be seen that Knapp's method is more conservative.

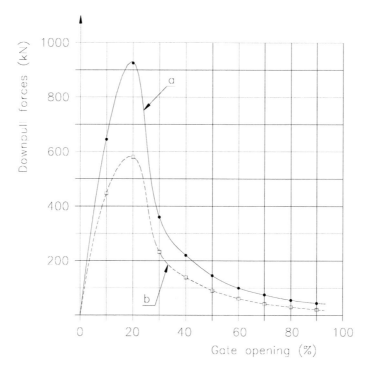

Fig. 8.27 Downpull forces. (a) Example 8.2; (b) Example 8.1

REFERENCES

1. Knapp, F.H., *Ausfluss, Überfall und Durchfluss im Wasserbau*, Verlag G.Braun, Karlsruhe (1960).
2. Murray, R.I. and Simmons, W.P., Hydraulic Downpull Forces on Large Gates, U. S. Dept. of the Interior, Bureau of Reclamation, *Research Report No.4* (1966).
3. Sagar, B.T.A., Downpull in High-Head Gate Installations, *International Water Power and Dam Construction* (March 1977).
4. Sagar, B.T.A. and Tullis, J.P., Downpull on Vertical Lift Gates, *International Water Power and Dam Construction* (Dec. 1979).
5. U. S. ARMY CORPS OF ENGINEERS, *Hydraulic Design of Reservoir Outlet Structures*, Manual EM 1110-2-1602 (1963).
6. Naudascher, E., Kobus, H.E. and Rao, R.P., Hydrodynamic Analysis for High-Head Leaf Gates, Paper No. 3904, *Journal of the Hydraulics Division, Proceedings of ASCE*, Vol. 90, No. HY3 (May 1964).
7. U. S. ARMY CORPS OF ENGINEERS, Hydraulic Design Criteria, *Hydraulic Design Chart 320-1*, Control Gates, Discharge Coefficients (1961).

Chapter 9

Gate Operating Forces

9.1 INTRODUCTION

The gate hoist is designed to overcome the resistant forces arising during the gate movement. In general, the following forces are to be considered:

G = gate weight
E = buoyancy of the submerged part
F_a = friction forces on supports
F_v = friction forces on seals
F_h = downpull (or uplift) forces.

For rotating gates (segment and flap, for example), the hoist capacity is determined by means of the torque exerted by the resistant forces about the rotation axis.

9.2 GATE WEIGHT

The components of the gate weight are:
a) weight of the gate structure;
b) weight of accessories and mechanical parts (wheels, pins, seals and so on) attached to it;
c) weight of paint;
d) weight of debris eventually retained in the gate structure;
e) weight of water eventually retained in the gate structure, in case of weir gates;
f) weight of ballast, if any.

The portions *c*, *d* and *e* should be considered only when acting unfavorably. For determination of the weight of paint and debris, the weight of the gate structure can be multiplied by 1.05.

For calculation of the weight of the gate structure, the following mean values of specific mass can be adopted:

Material	Specific mass (kg/dm^3)
Steel	7.85
Bronze	8.80
Natural rubber	0.93
Synthetic rubber	1.23
Wood	1.02

9.3 FRICTION ON SUPPORTS AND HINGES

The friction on supports and hinges is proportional to the water load on the gate and to the friction coefficient of the surfaces in contact. It is calculated by the classical equation

$$F_a = \mu N \qquad\qquad (9.1)$$

where

μ = friction coefficient
N = normal force on the support or hinge.

In most cases, the normal force consists of the hydrostatic thrust plus the component of the gate weight. The Coulomb assumption of μ = constant is just a first approach. In gates standing idle for long periods, it is often observed that there are additional friction forces due to the adherence between the surfaces in contact. Therefore, distinct values of the friction coefficient should be taken for both static and dynamic conditions. The following friction coefficients are applied:

Table 9.1 Friction Coefficients (μ)

Materials	Static conditions	Dynamic conditions
Bronze on steel	0.30	0.18
Steel on steel	0.40	0.20
Steel on self-lubricating copper alloy	0.15	0.10
Steel on concrete	0.40	0.40
PTFE on steel	0.10	0.10
Timber on timber	1.10	1.10
Timber on steel (*)	from 0.45 to 0.55	from 0.45 to 0.55

(*) along the fibers, $\mu = 0.45$
across the fibers, $\mu = 0.55$

In the case of trunnions of segment gates with sloped arms, the influence of the friction forces on the thrust washers due to the component normal to the pier face should also be analyzed, as shown in Figure 9.1.

The total resisting torque on the gate hinge is

$$M_m = \mu \left(F_r \frac{d}{2} + F_n \frac{d^*}{2} \right) \qquad\qquad (9.2)$$

where

    μ = friction coefficient

    d = inside diameter of bushing

    d* = mean diameter of the thrust washer

    F_r = radial component of the hydraulic thrust on each bushing

    F_n = axial component of the hydraulic thrust on each bushing.

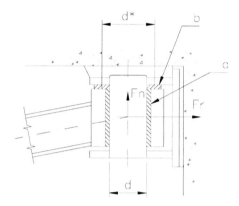

Fig. 9.1 Bearing reactions
(a) cylindrical bushing; (b) thrust washer

In segment gates with parallel arms, F_n is equal to zero and Equation 9.2 becomes

$$M_m = \mu \, F_r \, \frac{d}{2} \tag{9.2a}$$

The friction forces on the wheels are calculated by the formula

$$F_r = \frac{W}{R}\left(\mu\, r + f\right) \tag{9.3}$$

where

    F_r = friction force

    W = wheel load

    R = wheel radius

    r = wheel pin radius or mean radius of bearing

    μ = friction coefficient of bearing or bushing

    f = coefficient of rolling friction.

For flanged wheels a flange friction of 1/100 of the wheel load is to be considered when it acts unfavorably. Coefficients of friction for bronze bushings are shown in Table 9.1.

    A friction coefficient of 0.003 should be applied for roller bearings. According to the DIN 19704 standard, the coefficient of rolling friction f can be taken as equal to 0.05 cm for wheel tread hardness up to 150 BHN. For steels with hardness between 150 and 300

BHN, f is considered equal to 0.02 cm. Intermediate values may be determined by linear interpolation.

9.4 SEAL FRICTION

Seal friction forces are determined by the formula

$$F_v = \mu\,N \tag{9.4}$$

where

 μ = friction coefficient between seal and seat
 N = reaction force of the seat

The friction coefficient of rubber seals in contact with steel is equal to 1.0 for the starting and to 0.80 for the running condition. For rubber seals coated with a film of PTFE (Teflon, for example), the friction coefficient indicated in Table 9.1 may be used.

 The reaction force of the seat is calculated according to the type and geometry of the seal:

- J-seals

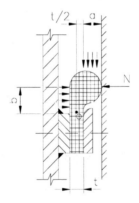

Fig. 9.2

Moments about point O:

$$p\,L\,b\frac{b}{2} + p\,L\,a\left(\frac{a}{2}+\frac{t}{2}\right) - N\,b = 0 \tag{9.5}$$

$$\therefore N = \frac{p\,L}{2\,b}(b^2 + a^2 + at) \tag{9.6}$$

where

 p = hydrostatic pressure on the seal
 L = seal length
 a, b, t = see Figure 9.2.

- Angle seals

Moments about point O:

$$p\,L\,b\,\frac{b}{2}\,-\,N\,a\,=\,0 \tag{9.7}$$

$$\therefore N\,=\,p\,L\,\frac{b^2}{2\,a} \tag{9.8}$$

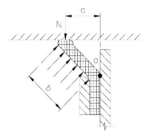

Fig. 9.3

Example 9.1 Submerged fixed-wheel gate

Determine the operating forces of the fixed-wheel gate of Example 8.1. The seal span is 5.5 m and the seal height is 7.9 m. The gate weighs 320 kN and is equipped with 36 cm diameter wheels assembled on roller bearings with effective diameter of 22 cm. The hardness of the wheel-rolling surface is 250 BHN. Side and top seals are J-shaped and arranged as per Figure 9.4.

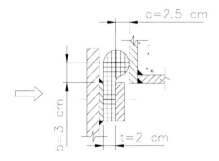

Fig. 9.4

Solution:

The total lifting load is given by the expression

$$C_A = (G - E) + F_r + F_v + F_h$$

where
  G = gate weight
  E = buoyancy
  F_r = wheel friction
  F_v = seal friction
  F_h = hydrodynamic forces

During the closure operation, the friction forces have the upward direction. Thus, the corresponding closure load is:

$$C_F = (G - E) - F_r - F_v + F_h$$

- Gate weight

  $$G = 320 \text{ kN}$$

- Buoyancy

  $$E = 320/7.85 = 40.8 \text{ kN}$$

- Wheel friction

Its maximum value occurs during the start of the opening operation and is calculated by Equation 9.3. Since the gate moves vertically, the resultant load on wheels is composed solely by the hydrostatic load acting on the gate. So,

$$W = \gamma \ B \ h \ (H - \frac{h}{2}) = 9.81 \cdot 10^{-6} \cdot 5500 \cdot 7900(25750 - \frac{7900}{2}) = 9292 \cdot 10^3 \text{N}$$

Also,
  R = 36/2 = 18 cm
  $\mu = 0.003$
  r = 22/2 = 11 cm
  f = 0.02 cm

Hence,

$$F_r \ = \ \frac{9292}{18} \ (0.003 \cdot 11 + 0.02) \ = \ 27.4 \text{ kN}$$

- Friction on side seals

Its maximum value occurs with the gate closed, at the start of opening. The gate has two side J-seals, each measuring 7.9 m long. Applying Equation 9.6, gives:

$$F_{VL} \ = \ 2\mu N \ = \ \mu p \frac{L}{b} \ (b^2 + a^2 + at)$$

Head on the middle length of side seals:

$$H = 25.75 - 7.9/2 = 21.8 \text{ m}$$

So, the water pressure on the side seals is

$$p = 21.8 \cdot 9.81 = 213.86 \text{ kN/m}^2$$

The friction coefficient between the rubber seal and the steel seat is 1.0 at the start and 0.8 in motion. Hence,

. at the start,

$$F_{VL} = 1.0 \cdot 213.86 \; \frac{7.9}{0.03} \; (0.03^2 + 0.025^2 + 0.025 \times 0.02) = 114 \text{ kN}$$

. in motion,
$$F'_{VL} = 0.8 \cdot 114 = 91.2 \text{ kN}$$

- Friction on top seal

The top seal is 5.5 m long and the head on this seal is

$$H = 25.75 - 7.9 = 17.85 \text{ m}$$

which corresponds to a water pressure of

$$p = 17.85 \cdot 9.81 = 175.11 \text{ kN/m}^2$$

At the start, the top seal friction is

$$F_{VS} = \mu \, N = \mu \, p \, \frac{L}{2b} \, (b^2 + a^2 + a \, t) =$$

$$= 1.0 \cdot 175.11 \frac{5.5}{2 \cdot 0.03} (0.03^2 + 0.025^2 + 0.025 \cdot 0.02) = 32.5 \text{ kN}$$

- Hydrodynamic forces: see solution of Example 8.1.

In the tabulation of the resisting forces, the following simplifying assumptions have been made, for ease of understanding:
a) wheel friction forces - uniform variation from the maximum value (gate closed) to zero (gate open);
b) side seals friction forces - the maximum value occurs at the start, with the gate closed. As soon as the gate starts to open, it reduces to the value corresponding to the gate in movement. From this point on, the friction forces reduce linearly until they become zero, with the gate totally open;

c) top seal friction force - occurs only when the top seal is in contact with the lintel and is taken equal to the maximum value corresponding to the starting condition, at 10 per cent gate opening.

Under these conditions, the following table is formed.

GATE OPENING (y/y_o x 100)	0	10	20	30	40	50	60	70	80	90	100
Weight less buoyancy (G - E)	279.2	279.2	279.2	279.2	279.2	279.2	279.2	279.2	279.2	279.2	279.2
Wheel friction (F_r)	27.4	27.1	21.9	19.2	16.4	13.7	10.9	8.2	5.5	2.7	-
Side seal friction (F_{VL})	(114) 91.2	82.1	73.0	63.8	54.7	45.6	36.5	27.4	18.2	9.1	-
Top seal friction (F_{VS})	32.5	32.5	-	-	-	-	-	-	-	-	-
Downpull force (F_h)	-	454.1	584.2	233.8	140.5	91.3	57.7	40.5	28.1	24.3	-
Lifting load	(453.1) 430.3	875	958.3	596	490.8	429.8	384.3	355.3	331	315.3	279.2
Closing load	(105.3) 128.1	591.6	768.5	430	348.6	311.2	289.5	284.1	283.6	291.7	279.2

The above-calculated forces are summarized and plotted in the Figure 9.5.

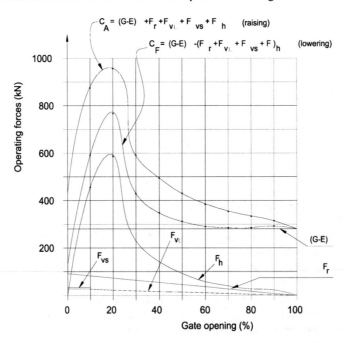

Fig. 9.5

Example 9.2 Spillway segment gate

Estimate the capacity of the two hoisting cylinders of the segment gate of Figure 9.6. The gate arms form an angle of eight degrees with the pier face and their hinges are equipped with self-lubricating bronze bushings with 40 cm internal diameter. The mean diameter of the thrust washers is 45 cm. The total water thrust on the gate is 23870 kN, the gate weighs 1280 kN and the horizontal distance between the gravity center and the gate hinge is 15.2 m. The side seals are made of Neoprene and have a unit length of 20 m.

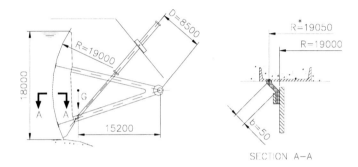

Fig. 9.6

Solution:

- Moment about the gate hinge due to the gate weight:

$M_p = 1280 \cdot 15.2 = 19456$ kN-m

- Moment due to the trunnion gate friction:

The reaction on each trunnion due to the water thrust has a radial component of

$F_r = 23870/2 = 11935$ kN

and an axial component (normal to pier face) of

$$F_n = \frac{23870}{2} \tan 8° = 1677.3 \text{ kN}$$

The friction coefficient of the self-lubricating bronze bushings is 0.15 at the start. Also,
d = 40 cm
d* = 45 cm

So, the moment in each pin will be (Eq. 9.2):

$$M_m = 0.15(11935\frac{0.4}{2} + 1677.3\frac{0.45}{2}) = 414.7 \text{ kN} - m$$

- Moment due to side seals friction:

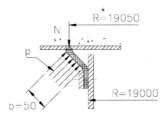

Fig. 9.7

Head at middle length of the side seals:
$\quad$ H = 18/2 = 9 m,
which corresponds to a water pressure of
$\quad$ p = 9 · 9.81 = 88.3 kN/m^2

Equation 9.8 gives the seat reaction on each side seal:
$$N = p \ L \ \frac{b^2}{2 \ a} = 88.3 \cdot 20 \ \frac{0.05^2}{2 \cdot 0.05} = 44.15 \ \ kN$$

The resisting moment on each side seal is, therefore:
$\quad$ $M_v = \mu \ N \ R^* = 1.0 \cdot 44.15 \cdot 19.05 = 841.1$ kN-m

The total resisting moment is
$\quad$ $M_r = M_p + 2 \ M_m + 2 \ M_v =$
$\quad\quad = 19456 + 2 \cdot 414.7 + 2 \cdot 841.1 = 21967.6$ kN-m

The torque exerted by the pair of hydraulic cylinders is
$\quad$ $M_s = 2 \ C \ D$

where
$\quad$ C = capacity of each hydraulic cylinder
$\quad$ D = distance between gate hinge and cylinder (see Figure 9.6).
$$\therefore C \ = \ \frac{M_s}{2 \ D}$$

Taking $M_s = M_r$,

$$C \ = \ \frac{21967.6}{2 \cdot 8.5} \ = \ 1292.2 \ kN$$

Considering a safety margin of 20 per cent in the capacity of the gate actuators, the rated capacity of each hydraulic cylinder should be, at least,

$\quad$ C = 1.2 · 1292.2 = 1550.7 kN.

Chapter 10

Aeration

10.1 INTRODUCTION

In gated conduits that do not discharge directly into the atmosphere, such as intakes and some bottom outlets, a high velocity flow occurs downstream of the gate resulting in sub-atmospheric pressures. This drop in pressure, which is dependent on the flow rate, the gate opening and the geometry of the installation, theoretically can be as low as the vapor pressure of water and may disturb the gate operation, due to the occurrence of vibration and cavitation.

In the earlier designs of high-head gates there was no provision for air vents nor were precautions taken to keep the walls of the water passages free from cavities or projections, resulting in severe damages to the gates and tunnels. Among the first gates thus designed are those installed in the Roosevelt Dam, USA, in 1908 [1]. These gates, 1.5 m wide by 3 m high, were intended for service under a maximum head of 67 m. The gates were put into service while the reservoir was still bellow its maximum level, but trouble soon developed such as concrete erosion and loosening of bolts, fastenings and metal linings. After the damage had been repaired, the gates were again put into service. As the problems returned, the users decided for their abandonment.

In 1909, four slide gates 1.1 m wide by 1.96 m high were installed in the Pathfinder Dam, USA [1]. In operation, hammering and reverberating sounds were heard. The intensity of these increased as the flow through the gates increased, until they resembled a thunderstorm and blasts at maximum discharge, causing the dam to shake. After closure of the conduit, large masses of loose rock and portions of concrete below the damaged gate were found; the 19 mm steel lining was torn as if it had been made of paper. An air shaft was cut through the roof immediately downstream from the gates and the tunnel was repaired. The gates were put into service again and the solution proved quite effective. Ever since, high-head gates have been provided with an air vent designed to admit air in large quantities into the water passages closely adjacent to the downstream leaf face and to keep the pressure near to atmospheric.

10.2 AIR VENT - FUNCTIONS AND FEATURES

Air vents are indispensable in high-head installations of gates with a downstream skin plate and seals. On the other hand, there is no need of air vents for gates with an upstream skin plate and seals, since the air demand can be provided through the gate shaft. Special care should be taken when designing the air vent for bottom outlet segment gates since it is not always possible to provide a free passage in the vertical shaft so as to permit adequate aeration. This is the case, for example, of designs where an operation platform is built just above the gate. This platform hinders air circulation and requires the construction of an air vent outlet in one of the sidewalls of the gate chamber.

An interesting case occurred in Brazil due to the alteration introduced in the original design of the air vent of a bottom outlet segment gate with 3.1 m span and 3.4 m high, operating under 55 m head. Originally, the design contemplated a 1.2 m diameter pipe with an outlet in one of the sidewalls of the gate chamber. Yet the outlet was installed near the top of the wall located upstream of the gate (see Figure 10.1).

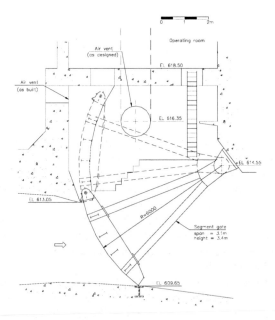

Fig. 10.1 Bottom outlet segment gate

Therefore, by opening the gate, the air vent outlet was gradually closed by the skin plate up to its almost total isolation. Aeration was therefore ineffective, causing strong vibrations to the tower. The depression created downstream of the gate drained part of the air from the nearby region with which it communicated through openings for the passage of the hoist stems and of the maintenance staff, that is, from the tower interior, creating great discomfort for the persons present in the gate operating floor. The problem was partly solved by limiting the maximum gate opening and the construction of walls and doors for separation of the operating room from the depression area.

The primary functions of the air vent are:

a) reduce or eliminate sub-atmospheric pressure in the conduit during emergency closure or partial-gate operation;

b) permit drainage of the conduit; and

c) allow air to escape when the conduit is being filled.

The air vent is usually formed in the concrete structure. In some cases, a metal pipe is embedded in the concrete, thus eliminating the molds. Most vents are circular in shape. In some projects, a square or rectangular section is chosen to facilitate the mold execution. For greater effectiveness, the lower end of the air vent should be located in the conduit ceiling, as close to the gate as possible, at a distance not over 2 m. The vent inlet should be located sufficiently above the maximum reservoir water level, preferably on the downstream face of the intake, to avoid interference with the air-flow. The air vent should be as straight as possible, with the minimum number of bends and sharp corners, and without abrupt changes in section.

10.3 AIR VENTS - EMPIRICAL CALCULATION

Sarkaria and Hom [2] suggested the following empirical equation for solution of the vent diameter size for closed conduit systems without surge tanks:

$$d = 0.291 \left(\frac{p^2 L}{H_n^2} \right)^{0.273} \qquad (10.1)$$

where
  d = diameter of the air vent, in m
  P = rated output of the turbine, in MW
  L = length of the air vent, in m
  H_n = rated head of the turbine, in m.

The authors point out that, as a precaution, the air vent size calculated according to the above formula should be considered as a minimum recommendation. In addition, in cold climates, where there is the possibility of ice formation in the air vent, a greater value should be adopted to counterbalance the effective area reduction.

Example 10.1 Determine the diameter of the air vent for an intake gate of a 350 MW turbine, rated head of 80 m. The length of the air vent is 50 m.

Solution:

  $P = 350$ MW
  $L = 50$ m
  $H_n = 80$ m

$$d = 0.291 \left(\frac{350^2 \cdot 50}{80^2} \right)^{0.273} = 1.895m \rightarrow \text{adopted } d = 1.9 \text{ m}$$

10.4 AIR-DEMAND RATIO

The great interest in determining the air demand induced various authors to conduct systematic research and prototype measurements. The results brought about a great dispersion of values and various formulas were developed for calculating the air-demand ratio, which is given by

$$\beta = \frac{Q_a}{Q_w} \tag{10.2}$$

where
Q_a = flow rate of air
Q_w = water discharge

The β ratio is a function of various parameters such as the conduit and gate geometry, the velocity and depth of the vena contracta and the water head. Most published papers suggests the following formula for determination of the air-demand ratio:

$$\beta = K \left(F_c - 1 \right)^n \tag{10.3a}$$

where
F_c = Froude number at vena contracta
K and n = empirical coefficients.

The Froude number is defined by

$$F_c = \frac{V_c}{\sqrt{g\,h_c}} = \frac{\sqrt{2\,g\,H}}{\sqrt{g\,h_c}} = \sqrt{\frac{2\,H}{h_c}} \tag{10.4}$$

where (see Figure 10.2)
V_c = velocity of water at vena contracta
h_c = depth of water at vena contracta
H = effective head at vena contracta.

Outstanding among the works are:

- Campbell and Guyton (1953)

$$\beta = 0.04 \left(F_c - 1 \right)^{0.85} \tag{10.3b}$$

The authors point out that maximum air demand takes place with the gate 80 per cent open and recommend that the maximum air velocity in the air vent should be limited to 45 m/s [3].

- U. S. Army Corps of Engineers (1964)

$$\beta = 0.03 \, (F_c - 1)^{1.06} \tag{10.3c}$$

Maximum air demand occurs at a gate opening of 80 per cent in sluices and a gate lip with a 45-degree angle on the bottom can be expected to have a contraction coefficient of approximately 0.80. The Froude number should be based on the effective depth at the vena contracta. Thus,

$$h_c = 0.8 \cdot 0.8 \, h = 0.64 \, h$$

where h is the height or maximum opening of the gate. Maximum air velocity is also limited to 45 m/s [4].

- Levin (1965)

$$\beta = K \, (F_c - 1) \tag{10.5}$$

The K coefficient is taken according to Table 10.1 [5].

Table 10.1 K-Coefficient

Conditions	K
Vertical lift gate in circular tunnel (Fig. 10.2a)	0.025 to 0.04
Same as above, with progressive transition from circular to rectangular section, followed by a very progressive transition (invert angle with horizontal lower than 10°) to circular section (Fig. 10.2b)	0.04 to 0.06
Same as above, with fast transition from circular to rectangular section, and from rectangular to circular section (Fig. 10.2c)	0.08 to 0.12

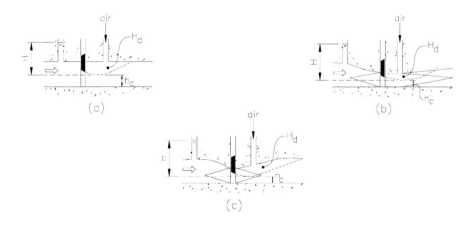

Fig. 10.2 Tunnel gates (Levin)

- Sharma (1976)

Studies carried out by Sharma [6] at the Technical University of Norway, Trondheim, Norway, correlate air demand and type of flow downstream of the gate. The research lists six types of flow (see Figure 10.3).

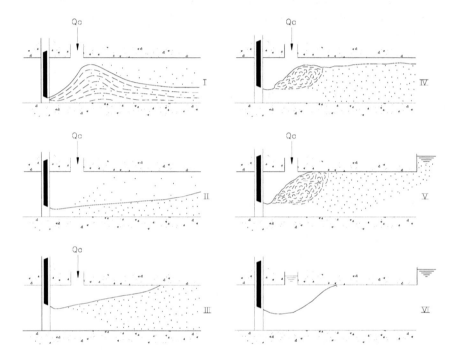

Fig. 10.3 Flow in tunnel gates (Sharma)

The first four types correspond to free surface flow, with air-entrainment in the flow. Observations in prototypes have shown that in these cases the air demand depends on the Froude number in the vena contracta and on the ratio of the length of the conduit downstream the gate (L_g) and the height or diameter of the conduit (D_g).

For free flowing conduits, Sharma developed the Figure 10.4, which shows a series of curves based on experimental data, in the form of $(1 + \beta)$ versus (A_c/A_g) for different values of Froude numbers, where

A_c is the area of flow at vena contracta

A_g is the cross-sectional area of the conduit.

The plotted curves were based on specific values of L_g/D_g. Therefore, the values obtained from the Figure 10.4 cannot be considered very accurate - only approximate or general.

The upper curve represents the equation

$$1 + \beta = \frac{1}{A_c/A_g} \qquad\qquad (10.6)$$

which corresponds to the maximum possible air transport, where the Froude number at the vena contracta is high enough to create foamy flow and the depth of flow at the vena contracta is comparable to the vertical dimension of the conduit. In practice these two conditions are rarely satisfied.

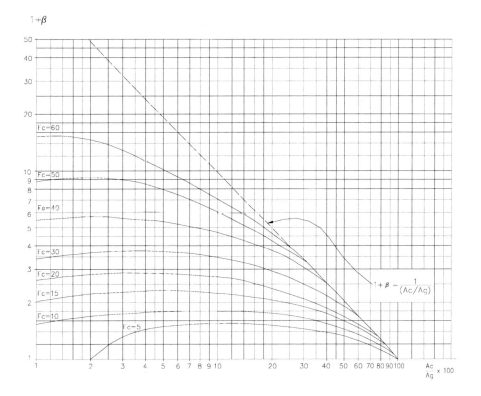

Fig. 10.4

The hydraulic jump followed by pipe flow (Fig. 10.3, type V) occurs in bottom outlets with the lower end submerged and a low relation of L_g/D_g, or during emergency closure of intake gates. For these cases, Sharma suggests the equation of Kalinske and Robertson with replacement of the Froude number at the jump location by the Froude number at the vena contracta. The expression then becomes:

$$\beta = 0.0066 \ (F_c-1)^{1.4} \qquad\qquad\qquad (10.3d)$$

In the type VI flow there is no air demand.

Figure 10.5 shows a comparison between the various calculation formulas of the air-demand ratios, for intake leaf gates with downstream seals and skin plate, or bottom outlets with lower end drowned (type V flow, according to Sharma). A coefficient K equal to 0.04 was used in the Levin's curve.

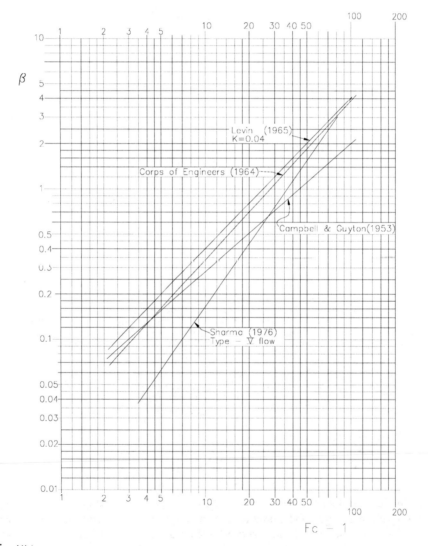

Fig. 10.5

10.5 AIR VENT DIMENSIONING

According to Levin, the air-flow in the air vent is

$$Q_a = 28\, m_a\, S_a\, \sqrt{2\, g\, H_d} \qquad\qquad (10.7)$$

where:
   m_a = flow coefficient of air vent
   S_a = cross-sectional area of air vent
   H_d = depression downstream of the gate (see Figure 10.2)

The water discharge, in turn, is given by

$$Q_w = B_c\, h_c\, \sqrt{2\, g\, H} \qquad\qquad (10.8)$$

where:

   B_c and h_c = width and depth of water at vena contracta
   H = effective head at vena contracta (see Figure 10.2)

Since

$$Q_a = \beta\, Q_w \qquad\qquad (10.9)$$

we have

$$28\, m_a\, S_a\, \sqrt{2\, g\, H_d} = \beta\, B_c\, h_c\, \sqrt{2\, g\, H} \qquad\qquad (10.10)$$

$$\therefore S_a = \frac{\beta\, B_c\, h_c}{28\, m_a} \sqrt{\frac{H}{H_d}} \qquad\qquad (10.11)$$

According to Levin, the flow coefficient of the air vent is calculated by

$$m_a = \frac{1}{\sqrt{\Sigma C_o + \lambda\, \dfrac{L}{d}}} \qquad\qquad (10.12)$$

where:
   $\Sigma\, C_o$ = sum of loss coefficients of obstacles such as entrances, exits, elbows, curves and
        screens
   λ = friction loss coefficient
   L and d = length and diameter of the air vent

The friction coefficient in the air vent may be accurately determined in the Moody chart, as a function of the Reynolds number and the relative roughness $\varepsilon\, /d$.

The Reynolds number is given by

$$R_e = \frac{V_a\,d}{v}$$

(10.13)

in which
V_a = air velocity, in cm/s
d = air vent diameter, in cm
v = kinematic viscosity of air, in cm^2/s.

The kinematic viscosity of air at a temperature of 20° C can be taken equal to 0.17 cm^2/s. For the roughness ε,

Material	ε (cm)
Concrete	0.03
Galvanized steel	0.015
Carbon steel	0.005

The following friction loss coefficients may be used as average values:

Concrete	$\lambda = 0.015$
Steel	$\lambda = 0.012$

The head loss coefficients C_o may be calculated in accordance with Tables 10.2, 10.3 and 10.4 developed by ASHRAE [7] for circular air ducts. The equivalence between circular and rectangular or square sections is given in Equations 10.16 and 10.7, respectively.

a) Entrances

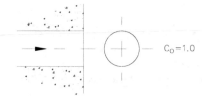

$C_o = 0.5$

Fig. 10.6

b) Exits

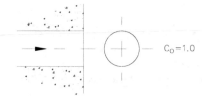

$C_o = 1.0$

Fig. 10.7

c) Screens

Let

$$n = \frac{S_g}{S_a}$$ (10.14)

where:
  S_g = net area of screen
  S_a = cross-sectional area of the air vent

Table 10.2 Local Loss Coefficients, Screens (ASHRAE)

n	0.30	0.40	0.50	0.55	0.60	0.65	0.70	0.75	0.80	0.90	1.00
C_o	6.20	3.00	1.70	1.30	0.97	0.75	0.58	0.44	0.32	0.14	0

For screens installed in entrances or exits, the C_o coefficient indicated in this item should be added to the respective coefficient (0.5 or 1.0 as the case may be).

d) Elbows

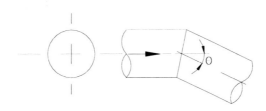

Fig. 10.8

Table 10.3 Local Loss Coefficients, Mitered Elbows (ASHRAE)

θ	20	30	45	60	75	90
C_o	0.08	0.16	0.34	0.55	0.81	1.20

e) Elbows with smooth radius

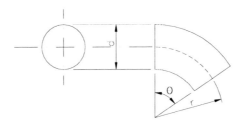

Fig. 10.9

Table 10.4 Local Loss Coefficients, Elbows with Smooth Radius (ASHRAE)

	θ	r/d					
		0.50	0.75	1.00	1.50	2.00	2.50
	30	0.32	0.15	0.10	0.07	0.06	0.05
C_o	45	0.43	0.2	0.13	0.09	0.08	0.07
	60	0.55	0.26	0.17	0.12	0.10	0.09
	90	0.71	0.33	0.22	0.15	0.13	0.12

The determination of the flow coefficient of the air vent is usually carried out by trial and error, starting with the knowledge of the tube geometry and of a value estimated for the diameter. With these, head losses and the cross section area are determined. Then, the diameter is checked; if it differs from the value initially attributed to it, another value is selected and the above calculation is repeated.

Air velocity V_a in the air vent depends directly on the depression H_d. For a Mach number below 0.5, we have:

$$V_a = 28\, m_a\, \sqrt{2\, g\, H_d} \qquad\qquad (10.15)$$

The depression H_d formed downstream of the gate should not exceed certain limits. The larger its value, the greater the likelihood of the occurrence of pulsation and cavitation. Levin states that depressions corresponding up to a head of 1.5 m can be tolerated without problem in well designed installations.

The U. S. Army Corps of Engineers suggests limiting the air velocity to 45 m/s which, for an air-flow coefficient of 0.7, is equivalent to a depression of about 0.3 m head. In Japan, the air vents are designed to keep the air velocity below 90 m/s [8].

Air vents with rectangular cross-section are calculated, as if the sections were circular. The equivalence of the circular and rectangular sections, for pipes of the same length, flow and head losses, is given by the following ASHRAE formula:

$$D_e = 1.38 \sqrt{\frac{(a\,b)^5}{(a+b)^2}} = 1.3\, \frac{(a\,b)^{0.625}}{(a+b)^{0.25}} \qquad\qquad (10.16)$$

where:
D_e = diameter of the equivalent circular section
a and b = rectangle dimensions.

In the particular case of a square section, the equivalence is given by:

$$D_e = 1.093\, a \qquad\qquad (10.17)$$

where a is the length of the square side.

Example 10.2 Design the air vent for an intake fixed-wheel gate 7.6 m wide by 12 m high, subjected to a head of 58 m. The air vent has a square section and is formed in the concrete structure of the intake, with a 90° elbow and a total length of 50 m. A screen that obstructs about 10 per cent of the total cross-sectional area of the air vent protects the exit.

Solution:

The maximum air-flow takes place with the gate 80 per cent open and the contraction coefficient of the jet issuing underneath the gate is 0.8. Therefore, the Froude number is calculated for a water depth h_c in the contracted section equal to

$$h_c = 0.8 \cdot 0.8 \cdot 12 = 7.68 \text{ m}$$

Hence, $H = 58 - 7.68 = 50.32$ m

- Froude number: according to Equation 10.4,

$$F_c = \sqrt{\frac{2\,H}{h_c}} = \sqrt{\frac{2 \cdot 50.32}{7.68}} = 3.62$$

- Air demand ratio: the Corps of Engineers' formula gives:

$$\beta = 0.03 \ (F_c-1)^{1.06} = 0.03 \ (3.62-1)^{1.06} = 0.083$$

First try: $d = 1.2$ m (initially the air vent will be calculated as having circular section). The local head losses are:

- in the entrance, $C_o = 0.5$

- in the exit, $C_o = 1.0$

- in the screen (n = 0.90), $C_o = 0.14$

- in the elbow, $\underline{C_o = 1.2}$
                                              $\Sigma C_o = 2.84$

Since the air vent is made of concrete, assume $\lambda = 0.015$; thus,

$$m_a = \frac{1}{\sqrt{2.84 + 0.015 \dfrac{50}{1.2}}} = 0.537$$

Considering a maximum depression downstream of the gate $H_d = 1$ m of head, the air vent cross-section area is:

$$S_a = \frac{\beta \ B_c \ h_c}{28 \ m_a} \sqrt{\frac{H}{H_d}} = \frac{0.083 \cdot 7.6 \cdot 7.68}{28 \cdot 0.537} \sqrt{\frac{50.32}{1}} = 2.29 \text{ m}^2$$

which corresponds to a diameter $d = 1.71$ m > 1.2 m.

Second try: $d = 1.7$ m.

- Air demand ratio:

$$m_a = \frac{1}{\sqrt{2.84 + 0.015\dfrac{50}{1.7}}} = 0.552$$

- Cross-sectional area:

$$S_a = \frac{0.083 \cdot 7.6 \cdot 7.68}{28 \cdot 0.552}\sqrt{\frac{50.32}{1}} = 2.22 \text{ m}^2$$

corresponding to a diameter $d = 1.68$ m, which confirms the selected value ($d = 1.7$ m).
The air flow is

$$Q_a = 28\, m_a\, S_a\, \sqrt{2\, g\, H_d} = 28 \cdot 0.552 \cdot 2.22\, \sqrt{2 \cdot 9.81 \cdot 1} = 152 \text{ m}^3/\text{s}$$

and the air velocity is

$$v_a = Q_a / S_a = 152/2.22 = 68.5 \text{ m/s}$$

The equivalence of the circular and square sections is

$D_e = 1.093\, a$
$D_e = d = 1.7 = 1.093\, a$
$\therefore a = 1.7/1.093 = 1.56$ m.

REFERENCES

1. Davis, C.V. and Sorensen, K.E.: *Handbook of Applied Hydraulics*, 2nd edition, McGraw-Hill.
2. Sarkaria, G.S. and Hom, O.S.: Quick Design of Air Vents for Power Intakes, *Proceedings of ASCE, Journal of the Power Division*, PO6 (Dec. 1959).
3. Campbell, F.B. and Guyton, B.: Air Demand in Gated Conduits, *IAHR Symposium* (1953), Minneapolis.
4. U. S. ARMY CORPS OF ENGINEERS: *Hydraulic Design Criteria*, Air Demand, Regulated Outlet Works, Sheet 050-1 (1964).
5. Levin, L.: Calcul Hydraulique des Conduits d'Aeration des Vidanges de Fond et Dispositifs Deversants, *La Houille Blanche*, No.2 (1965).
6. Sharma, H.R.: Air-Entrainment in High-Head Gated Conduits, *Proceedings of ASCE, Journal of the Hydraulics Division*, HY11 (Nov. 1976).
7. ASHRAE-American Society of Heating, Refrigerating and Air Conditioning Engineers, Inc.: *ASHRAE Handbook* (1981) Fundamentals Volume.
8. Fujimoto, S, and Takasu, S.: Historical Development of Large Capacity Outlets for Flood Control in Japan, *XIII ICOLD Congress* (1979), New Delhi.

Chapter 11

Gate Hoists

11.1 INTRODUCTION

Gates can be operated by the action of the reservoir pressure, by a fixed hoist or by a moving device. The first case includes the drum gates, bear-trap gates and sector gates, which do not have hoists and only require valves and inlet and discharge piping in the recess chamber. The fixed-type hoists use screw lifts, wire ropes, roller chains and hydraulic cylinders for operation of the gate and will be discussed in detail in this chapter. Movable hoisting devices are used mainly for operation of stoplogs and diversion gates. They also are used in the maintenance of the gate leaf and operating devices of spillway, intake and draft tube main gates. Among the more common are the hand or electric traveling hoists, overhead cranes, gantry cranes and wheel or tractor-mounted cranes.

The main function of a gate-operating device is to develop a large operating force with low power supply. The force is required to overcome the weight of the moving parts, the friction forces, the hydrodynamic forces and occasional or accidental loads. As a rule, the lifting height is small, it being sufficient to displace the gate to a position that no longer obstructs the free passage of the water. In spillway gates, the minimum lifting is defined as a function of the hydraulic profile determined by the maximum design flow, and the hoist must be able to lift the gate to a height where the gate bottom seal stays about 1 m above the water surface.

Electricity and internal combustion engines are the usual power sources for operation of gates. However, the latter as a rule have their use confined to emergency situations. The operating systems should, preferably, be provided with devices which allows for hand lifting of the gate by an operator in the event of power failure.

11.2 SCREW LIFTS

The hand operated screw lift has its use confined to small slide or fixed-wheel gates subjected to low head, and usually comprises a vertical threaded stem connected to the gate leaf and prevented from rotating, a stem lift nut and a mechanism that permits nut rotation in the desired direction. The nut may be operated directly by a hand wheel, or indirectly, by means of bevel or worm gear systems (Fig. 11.1).

Fig. 11.1 Hand-operated screw-lift hoist

The nut has also a strut function and is usually supported on roller bearings. It has to be designed to support the gate and stem weights and the friction forces. In general, nuts are made of aluminum bronze, which presents good mechanical strength and a low friction coefficient. This mechanism is normally installed within a frame and mounted on a cast-iron pedestal screwed to the operating deck.

The stem is subjected to tensile stresses during lifting and to compressive stresses during lowering (in case of slide gates), and should be designed as a column. The stems can be made of SAE 4140 alloy steel or AISI 416 stainless steel and are provided with square or ACME threads, with lead equal to the pitch.

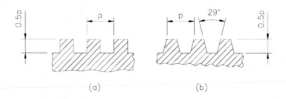

Fig. 11.2 Shapes of (a) square and (b) ACME threads

The screw lift may be driven by electric motors. Its operation is reliable, yet presents low efficiency and high cost when compared with wire rope hoists. Being self-locking, it cannot be used in installations where the gate closure is required under its own weight in the event of power failure. In motor driven screw lifts, a load limiter is provided (motor torque limiter, for example) to avoid excessive compression on the stem during the gate closure. This system can be further completed with a position indicator and limit switches corresponding to the gate travel extreme positions. A complete analysis of the stem and nut dimensioning is shown in the reference [1].

11.3 WIRE ROPES

Steel wire rope hoists are used exclusively in facilities where the gate closure is made by its own weight, for the cables cannot exert downward forces. Their use is very common in weir gates (particularly segment and fixed-wheel gates) in which the friction forces to be overcome are always less than the gate weight. In the basic design, it has two drums provided with two helical grooves (right and left-hand) where the steel cables connected to the gate leaf are coiled. The drums are installed one on each side of the gate and operated by pairs of gears or speed reducers and one electric motor. The torque is transmitted from one drum to the other through a coupling shaft. In gates with small ratio width/height, the lifting can be performed by means of a single drum and a cable set connected to the gate top center.

In general, the electric motors selected are three-phase AC, squirrel-cage induction rotor, with maximum torque limited to 300 per cent of the rated torque. Electromagnetic or hydraulic thrustor spring-operated shoe brakes are used for holding the suspended gate. Brakes and motors are attached to the speed reducer input shaft. Brakes actuate automatically when the current is switched off and should allow hand liberation upon power failure. The braking torque is 150 per cent of the torque corresponding to full load motor torque. The gate lowering occurs by gravity and the lowering speed may be controlled by motor energization in the direction of the descent or by means of an automatic regulator (a fan brake, for example).

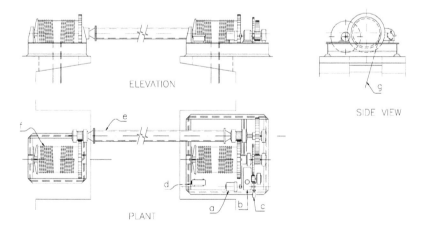

Fig. 11.3 Wire rope hoists for segment gate
(a) electric motor; (b) speed reducer; (c) brake; (d) position indicator; (e) coupling shaft; (f) drum; (g) wire rope

Drums are designed according to the selected cable. Their length must accommodate in a single layer the length of the cables required for the total lifting of the gate plus two holding turns. In addition, with the gate closed, two turns of the cables should remain coiled in the drum. Drum grooves are machined with the radius equal to 0.53 times the cable diameter. The minimum diameter of drums and pulleys is chosen in function the cable diameter and classification; it should follow the cable manufacturer specifications. As a rule, a minimum diameter equal to 25 times the cable diameter is adopted for drums and pulleys. In order to provide for rope run-off, the permissible fleet angle in relation to the pulley groove is usually limited to 3 degrees.

Material used in the manufacture of rope wire is stainless steel and galvanized or polished steel. The rope wires must present a minimum breaking strength in the range of 1570 to 1770 MPa, which corresponds to the American classification of plow steel. Ropes made of galvanized wires are used in certain applications where additional protection against rusting is required (contact with water, for example). However, their minimum effective breaking strength is 10 per cent less than ungalvanized wire rope. Stainless wire ropes are made of 18-8 steel and present great resistance to corrosion; notwithstanding their high cost, they are particularly indicated for submerged use. In Brazil, most gate hoist wire ropes are 6x19 and 6x37-class of construction. The construction is indicated by two numbers, the first giving the number of strands and the second being the number of wires in each strand. Various arrangements of wire (Seale, Filler and Warrington) are used in the construction of wire rope strands and do not affect the rope load capacity. When selecting the wire rope, the designer must take into account that the flexibility is inversely proportional to the wire diameter, whereas the abrasion resistance is directly proportional to that diameter. Thus, comparing 6x19 and 6x37 wire ropes of the same diameter, the first has greater resistance to abrasion, while the latter is more flexible.

The strands are wrapped in spirals about a central element, which is usually fiber or steel wire. Fiber core gives greater flexibility to the cable; steel core have high resistance to permanent deformation and increases the tensile strength. Ropes with steel core are rated 7.5 per cent stronger than those with fiber core [2]. Wire rope is made either preformed or non-preformed. In preformed ropes the wires and strands are pre-shaped into a helical form so that when laid to form the rope they tend to remain in place, with a minimum of internal stresses. Preformed ropes are more flexible and have greater fatigue strength.

Wire ropes are classified either regular lay or Lang lay, as illustrated in Figure 11.4.

Fig. 11.4 Lay of wire ropes - From left to right:
(a) right-regular lay; (b) left-regular lay; (c) right Lang lay; (d) left Lang lay

In regular lay, the wires in the strand are laid in the opposite direction to the lay of the strands in the rope. In Lang lay, the wires and strands are laid in the same direction. Lang lay ropes have greater flexibility than regular lay ropes and are more resistant to abrasion. Regular lay rope has fewer tendencies to spin. Lang lay ropes with fiber core should not be used as they present little stability and low resistance to permanent deformation [2]. The German DIN 19705 standard recommends the use of regular lay, heavily galvanized cables with a steel core covered with artificial fibers.

The rope ends are fastened by means of loops with clips or cast in sockets. The strength of a clip fastening is usually 80 per cent of the strength of the rope. When properly prepared, the strength of a socket fastening is approximately equal to that of the rope itself.

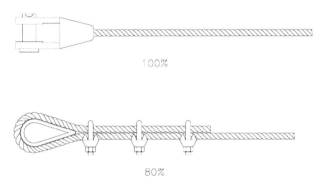

Fig. 11.5 Efficiency of fittings (CIMAF)

Figure 11.6 shows the correct application of clip fastening. Table 11.1 gives the proper spacing and number of clips for each size of wire rope.

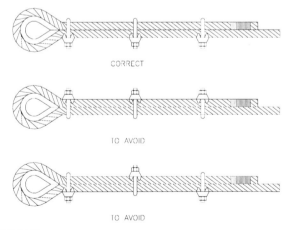

Fig. 11.6 Use of clips (CIMAF)

Table 11.1 Clips Required for Fastening Rope Ends (CIMAF)

Clip spacing (mm)	Rope diameter	Minimum no. of clips
29	3/16"	2
38	1/4"	2
48	5/16"	2
57	3/8"	2
67	7/16"	2
76	1/2"	3
95	5/8"	3
114	3/4"	4
133	7/8"	4
152	1"	4
172	1 1/8"	5
191	1 1/4"	5
210	1 3/8"	6
229	1 1/2"	6
248	1 5/8"	6
267	1 3/4"	7
305	2"	8
343	2 1/4"	8

A position indicator and top and bottom limit switches complete the cable hoist system. In certain cases, it is usual to install a device to detect rope loosening, as shown in Figure 11.7.

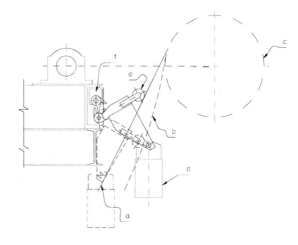

Fig. 11.7 Slack rope sensor
(a) wire rope; (b) loose wire rope; (c) drum; (d) counterweight; (e) sheave; (f) limit switch

In the event of rope loosening, the slack rope switch is operated by the action of the counterweight and interrupts the motor operation. Slacking of the ropes may be due to their rupture, to locking of the gate in an intermediate position or because the gate has reached its lowest position with the bottom limit switch ineffective.

11.4 ROLLER CHAINS

Design of chain hoists follows basically the same design principles of the wire rope hoists, where sprockets and chains replace drums and cables, respectively. This type of hoist exerts forces only when pulling the chains. For that reason, it can be used only in operation of gates whose closure is made by gravity. In vertical lift weir gates, the chains connect the hoist directly to the gate leaf top. When used to operate segment gates, the chains are installed downstream of the leaf; one end is fastened to the hoist frame or to the concrete, passes through a idler sprocket mounted on a fixed axle attached to the gate leaf bottom, through the driving sprocket and is stored in a special chain collector (Fig. 11.8).

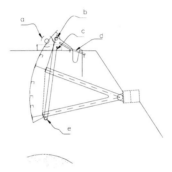

Fig. 11.8 Chain path
(a) hoist; (b) driving sprocket; (c) fixed end; (d) chain collector; (e) idle sprocket

The basic components of a chain hoist are: electric motor, brake, speed reducers, sprockets, chains, position indicator, limit switches, slack chain detector and hand drive.

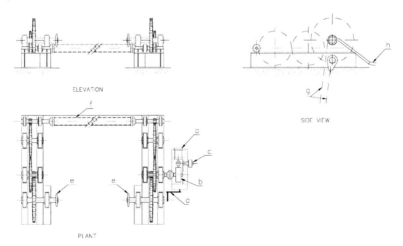

Fig. 11.9 Chain hoist
(a) electric motor; (b) speed reducer; (c) brake; (d) crank; (e) driving sprockets; (f) coupling shaft;
(g) chain; (h) chain collector

As a rule, the electric motors are three-phase AC, short-circuit or cage induction rotor, with maximum torque limited to 250 per cent of the rated torque. Electromagnetic or hydraulic thrustor spring-operated shoe brakes are used for holding the suspended gate.

Brakes actuate automatically when the current is switched off and should allow hand liberation upon power failure. The braking torque is 150 per cent of the torque corresponding to full load motor torque.

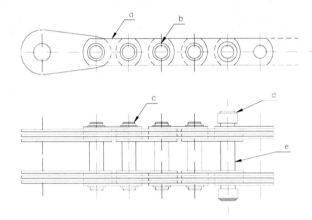

Fig. 11.10 Galle-type chain
(a) roller link; (b) retaining ring; (c) pin; (d) collectable roller; (e) long pin

Chains are of the Galle type and consist of plate manufactured link pairs, hinged in machined pins. The ends of the pins are provided with retaining rings. For storing purposes, elongated pins are installed at regular intervals. These pins are collected by the inclined guideways made from angle bars and reduce the chain length during the gate lifting.

SECTION A—A

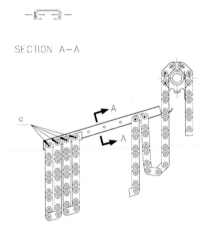

Fig. 11.11 Chain collector
(a) elongated pins

The chain driving pinions are either fitted on a shaft or made integral with the shaft. The sprockets have self-lubricating bushings fitted on stainless steel shafts. Pinions, sprockets and shafts are manufactured from chromium-molybdenum steels (series SAE 41XX) or nickel-chromium-molybdenum (series SAE 43XX), with an ultimate strength of about 1000 MPa. The teeth are always milled and should present a surface hardness of over 450 BHN.

The minimum number of teeth is 8 [3]. With z the number of teeth and p the chain pitch, the pitch circle diameter D_p can be determined as follows:

$$D_p = p/\sin(180°/z) \tag{11.1}$$

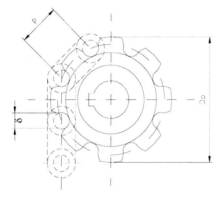

Fig. 11.12 Sprocket for roller chains

The following proportions are recommended for design of the chain links:

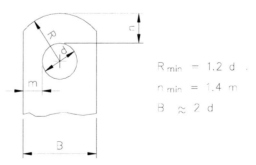

$R_{min} = 1.2\ d$

$n_{min} = 1.4\ m$

$B \approx 2\ d$

Fig. 11.13 Recommended dimensions for link plates

11.5 OIL HYDRAULIC DRIVES

One of the most commonly used gate driving mechanisms is the hydraulic hoist; it associates high load capacity with design simplicity, ease of control and operating reliability. It comprises a hydraulic cylinder operated by oil pressure supplied by a pumping unit.

Basically a hydraulic cylinder consists of a cylinder, upper and lower heads, a piston, and a piston rod (Fig. 11.14).

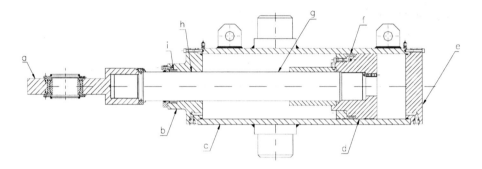

Fig. 11.14 Hydraulic cylinder (Rexroth Hydraudyne B.V.)
(a) clevis; (b) upper head; (c) cylinder shell; (d) piston; (e) bottom head; (f) piston seal; (g) stem;
(h) guiding strip; (i) rod seal

The hydraulic cylinder hoist may be either of single or double-action. The former is used only in gravity-closing gates. The double-action cylinder, used mainly in slide gates, pushes down the gate, overcoming all friction forces, which eliminates the need of counterweights. In this design, the stem and its extensions should be guided at intervals by intermediate bearings and checked against buckling. When compared to other types of hoist, the hydraulic hoist has the disadvantage of limited lifting height for, while the gate can be moved from the totally closed to the totally open position in a single operation, the complete withdraw of the gate or its lifting up to the maintenance deck requires either some external help or the gate lifting in partial operations with removal of the intermediate auxiliary stems.

Gates driven by cable or chain hoists can be totally lifted in a single operation. On the other hand, the hydraulic cylinder hoist offers several other advantages and favorable features, such as the possibility of exerting forces in both moving directions, greater mechanical efficiency and ease of speed control.

Lifting of the gate is carried out by oil pressure supplied by the pumping unit. Oil under pressure is directed to the interior of the cylinder, on the stem side, pushing up the piston. The upward movement of the piston forces the oil on the other side of the cylinder to return to the tank.

The closure of gates operated by single-action cylinders takes place under the weight of the gate and other moving components (piston and stem), which forces the transference of the oil under the piston to the upper part of the cylinder, with the help of the depression created there by the downward movement of the piston. Since the volume of the upper part is greater than that of the lower one (due to the stem), the difference is compensated by the oil contained in the tank, through the siphon effect. If the piping length is very big, or the tank is located below the top of the cylinder, the compensating volume of oil can be stored in a small additional tank (solution adopted for the intake gate hoists of the Itaipu Power Plant), or eventually supplied by the pumping unit itself. The oil pumped in excess in that operation returns to the tank.

Fig. 11.15 Hydraulic cylinders for the Itumbiara intake gates (KRUPP)

Fig. 11.16 Eye coupling for the hydraulic hoist stem of the Itumbiara intake gates (KRUPP)

Small diameter cylinders are manufactured from seamless steel pipes, whereas the large ones are fabricated from rolled plates longitudinally welded. The cylinder bore is machined and honed. Long cylinders are designed with the minimum possible number of transverse joints. Radiographic examination of all welds is required. Cylinders are required to pass a hydrostatic test of 150 per cent of the design pressure. Cylinder heads are made of cast or forged-steel although plates are sometimes used. In most cases, the requirements of the ASME-Unfired Pressure Vessel Code, Section VIII are followed in the design of cylinders. Pistons are made of bronze, gray cast iron or cast steel and are provided with circumferential split rings and also with a stuffing box with V-packing rings. Piston rods are manufactured from stainless steel, chrome-plated or ceramic coated laminated steels. The lower cylinder head is usually provided with a bronze sleeve stem guide and a wiper scraping to prevent entry of foreign matter from exposed stems. Sealing of flanges and transverse joints of cylinders are made with Neoprene O-rings. In most cases, the stems are designed with a slenderness ratio (L/R) less than 200.

The end positions of the gate travel are detected by limit switches designed to stop the operation in course. The hydraulic system is designed to keep the oil pressure under the piston with the gate totally open, for an undetermined period of time. If leakage occurs in the piping or in the piston rings, the gate starts to close. To prevent closure of the gate for that reason, it is customary to install in intake emergency gates hoisting systems, an additional limit switch located slightly below the fully raised position, which starts the pump and brings the gate back to its original maximum opening position.

To ensure easy seating of the gate on the sill at the end of the closure travel, a damping means (dash-pot) is provided in the bottom of the cylinder to reduce the area of the return oil pipe, thus restricting the return flow and, consequently, the descent speed.

Hydraulic cylinder hoists are driven either by individual hydraulic power units or jointly, by a single pumping unit.

Vane or gear pumps driven by electric motors are usually selected to operate hydraulic hoists. Pumps are commonly designed with a rated pressure 25 per cent greater than the design pressure and flow capacity about 10 per cent above the calculated one. Hydraulic hoists are usually designed for rated oil pressure from 7 to 25 MPa. The use of high pressures in cylinders of great capacity is very common and advantageous and permits reducing the cylinder diameter and therefore, its costs.

The basic components of a hydraulic power unit are:
. oil tank;
. filters;
. pumps;
. electric motors;
. flow-directing valves;
. pressure-relief valves;
. flow regulating valves;
. pressure switches;
. pressure gages;
. piping;
. control components (relays, push buttons etc).

The tank has the function of storing the oil used in the hydraulic system and its structure may be used for mounting the hydraulic control devices. There is no general rule to set the volume required by the tank: each case has to be studied separately. The minimum oil level must be such that the pump suction remains submerged about 100 mm so that air cannot be drawn in. Also, some manufacturers recommend locating the pump suction 50 mm far from the tank bottom so that sludge or any solid matter cannot be picked up. All return lines to the tank are always located below the minimum oil level and separated from the suction pipes by vertical bafflers to prevent the fluid coming back to the tank from returning immediately to the circuit without having performed an effective heat dissipation. Generally, the maximum temperature should be limited to 55 and 80-degree Centigrade inside the tank and the piping, respectively. In the case of frequent operation of the hydraulic system, there is a rise in the oil temperature and oil cooler should be installed in the tank. An alternative solution to this problem is to increase the oil volume in the tank.

Tanks are also provided with filling cap, oil level gauge, breather openings and drain plugs.

The pump suction filters are chosen according to the maximum flow and may be installed in parallel, if required. A rule of thumb is to select a filter that allows a maximum flow equal to three times the pump rated flow. The screen in the pump suction filter should be no coarser than 100 mesh. The filters in the pump discharge line should be able to remove particles above 0.01 mm and pass a flow equal to or greater than three times the maximum flow of the system. Filters are usually located inside the oil tank, which may hinder their removal for maintenance services. This can be avoided by installing the filters externally to the tank (see Figure 11.17).

Fig. 11.17 Hydraulic power unit for the Itumbiara intake gates (KRUPP)

An alternative solution has been adopted in some power plants of the former USSR, as quoted by Korotikov [4]. The method consists in the separation of the tank into two parts, one for the suction line and other for the pump discharge line, by means of a metal sheet provided in its lower portion with an opening covered with fine mesh. This element is introduced through a slit in the tank top and slides in vertical guides. The partition screen is easily removed and cleaned by compressed air.

The flow-directing valves are usually two-way solenoid operated valves, controlled or not by a pilot valve, depending on their function. The check valves, which are also directional valves, permit free fluid flow in one direction and prevent flow in the reverse direction. Pilot-controlled check valves are used to permit flow in the reverse direction beyond a determined point in the working cycle.

Safety devices are included in the oil circuit to protect the hydraulic system against eventual overpressure and consist of relief valves installed in the pump discharge line. The relief valve causes the oil to return to the tank when the pressure exceeds a preset value. Relief valves are usually set 10 per cent above maximum pressure, for the case of occasional load.

Flow regulating valves are installed in the line connected to the bottom of the cylinder and permit a very accurate control of the gate descent speed.

Figure 11.18 shows a schematic hydraulic diagram for two gravity-closing segment gates, each operated by two hydraulic cylinders, as those used in spillways. Two motor-driven oil pumps are provided, of which one is for reserve.

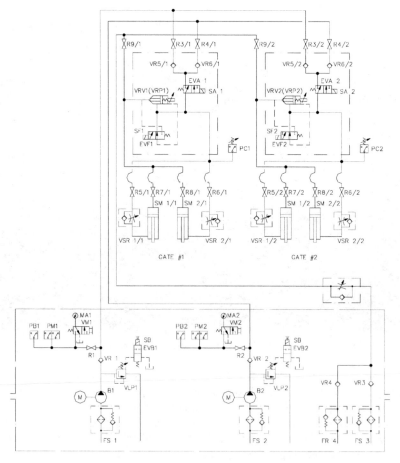

Fig. 11.18 Schematic hydraulic diagram for gravity-closing segment gates

The opening and closure operations are carried out with a gate in each turn and are described as follows.

- Gate opening:

Before starting, all solenoids (SA, SB and SF) are de-energized, their respective electrovalves EVA, EVB and EVF being out of action.
a) Once the opening command has been given, the pump motor and the EVA electrovalve are simultaneously energized;
b) The oil supplied by the pump enters the circuit through FS and B, returning to the tank through VLP and EVB;
c) After operating in idle for some seconds, the EVB valve is automatically energized, piloting VLP, which closes. In this situation, VLP discharges the oil in the tank whenever the pressure in the line exceeds the preset value;
d) The oil is pumped through VR1 (VR2), R3 (R4), VR5 (VR6), EVA, R5, R6, VSR1 and VSR2 to the underside of the piston in the cylinders SM1 and SM2;
e) The oil in the top of the cylinders returns to the tank through R7, R8, R9, VR7, VR4 and FR4;
f) The gate starts opening;
g) The opening operation is automatically interrupted when the programmed stop levels are reached, de-energizing therefore the pump motor and the EVA electrovalve;
h) A new opening command should be given upon every programmed stop, until total gate opening is attained;
i) In the closed position, EVA and EVF valves prevent the oil returning to the tank.

- Gate closure:

a) Once the closure command is given, EVA and EVF electrovalves are actuated and the pump motor starts operation;
b) The gate starts closure by gravity action and, automatically, after some seconds in idle, the EVB electrovalve is shifted;
c) The oil on the underside of the piston passes to the top of the cylinder through VSR1, VSR2, R5, R6, VRV, R7 and R8;
d) The pumped oil passes through VR1 (VR2), R3 (R4), VR5 (VR6), EVA and VRV. At the VRV outlet, the oil flow is divided and part of the oil passes through R7 and R8 to complete the oil volume in the top of the cylinder (compensation of the volumetric displacement of the cylinder stem). Another part of the oil returns to the tank passing through R9, VR7, VR4 and FR4;
e) The VRV valve regulates the oil flow and the rate of closure.

NOTE: The gate can be manually closed by shifting the EVF valve, in case of an emergency (power failure). The closure is made as described before, independent of the pump motor units. As this operation may cause air admission into the piping, air venting may be necessary. In this case, it is desirable to locate the oil tank at an elevation above that of the cylinders, in order to keep the piping always filled with oil.

Figure 11.19 shows an example of a hydraulic diagram for a gravity-closing gate operated by a hydraulic cylinder and a single pump motor unit. The system is provided with a hand pump for the event of power failure.

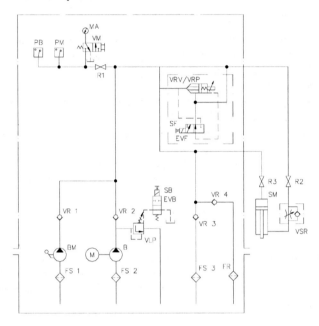

Fig. 11.19 Schematic hydraulic diagram for gravity-closing gate

Hydraulic circuits are also provided with the following safety devices:
. Filters provided with the safety valve to permit oil admission directly in the circuit, if an eventual clogging occurs;
. Pressure switches: once the opening command of gate no.1 has been given, the PB1 pressure switch can detect an eventual pressure drop in the circuit, commanding automatically the operation of the spare pump motor unit. If the pressure drop continues, the PB2 pressure switch de-energizes the spare pump motor. During the opening operation, the PM pressure switch may detect any overpressure, de-energizing the pump motor unit in operation;
. Flow-regulating valve with retention (VSR): if a pipe rupture occurs, the VSR regulates the oil outflow at a preset value, controlling the closure speed;
. PC pressure switches: if the gate locks during closure, pressure switch PC de-energizes the pump motor unit and takes the EVF valve out of action, stopping closure.

Piping should be as short as possible and sharp curves should be avoided. A minimum bend radius equal to three times the pipe diameter is recommended. Expansion joints should be provided where deformation in the pipe longitudinal direction is expected to occur.

The oil commonly used in the hydraulic systems is mineral oil, which should have good resistance to the formation of rust, oxidation and foam. Oils having a viscosity of 150 SSU at 38° C and a minimum viscosity index equal to 90 are used.

Pipes used in the hydraulic systems are usually black seamless steel pipe conforming to ASTM A53 or A106 standards. Pipe sizing is determined according to the maximum oil speed. Generally, the following speed limitations are observed in pipe design:

. in suction and filling lines: 1.2 m/s
. in return lines: 3.0 m/s
. in pressure lines:
  - continuous operation: 3.0 m/s
  - intermittent operation: 4.5 m/s
  - infrequent operation: 6.0 m/s.

11.6 GATE HOIST ARRANGEMENT

Hand or motor-driven screw lifts are vertically installed above the gate and mounted on the operating deck. When the design allows it, the threaded stem extends as a single piece from the gate leaf, where it is fastened to the lift nut by a pin. In facilities where the operating deck is distant from the gate, extension stems are used, supplied in sections with maximum length of 5 m and joined by means of screwed sleeves. In this case, intermediate bearings are installed with the double function of guiding the stems and reducing the buckling length for the gate closure operations. Bearings are installed on the dam wall face, with vertical spacing equal to or below 3 m.

The components of cable or chain gate hoists are assembled on metal frames manufactured from rolled sections and bolted onto the gate operating platform, or onto metal structures specially designed to permit complete withdrawal of the gate from its guides, by external means, with no need to displace the hoist. Water contact with the lifting cables or chains, due to immersion or from the discharge jet when the gate is open or from overleaf discharge, should be avoided whenever possible. Alternatively, the design may provide for galvanized or stainless steel cables or, if chains are used, stainless steel pins with bronze bushings. In the specific case of spillway segment gates, the location of cables upstream of the gate skin plate allows a substantial reduction in the hoist load capacity, for it exerts a greater lifting torque than would be obtained with downstream cables (see Figure 11.20).

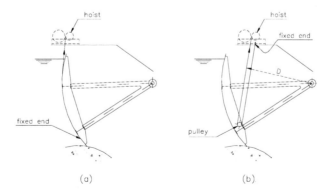

Fig. 11.20 Typical arrangements of wire rope hoists
(a) upstream wire ropes; (b) downstream wire ropes

Hydraulic cylinders can work submerged or not and in any position. When used to drive vertical lift gates in intakes or bottom outlets, they can be installed on the operating deck (supported on the cylinder bottom head); within the shaft (suspended by the upper head); or placed within the gate in a reverse position, that is, with a fixed stem and movable cylinder (see Figure 11.21).

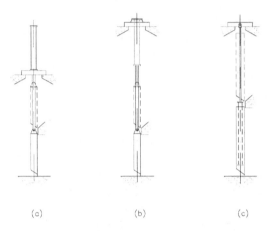

(a)   (b)   (c)

Fig. 11.21 Vertical lift gates – Arrangement of hydraulic hoist
(a) supported; (b) suspended; (c) embedded

A pair of hydraulic cylinders usually operates wide spillway gates. The cylinder stems are connected either to the lower part of the gate leaf or to an intermediate point in the radial arms. Cylinders are suspended in the piers either by the upper head or by pivots placed at half-length.

During motion of the segment gate, the cylinders rotate on the suspension axis. Therefore, the design should provide that feeding and return of the cylinder oil is carried out through rotating connections or flexible hoses. Pressure and return piping of both cylinders are installed symmetrically in relation to the vertical plane that passes through the span center, in order to equalize the head losses.

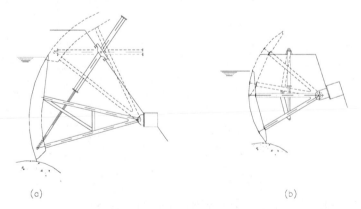

(a)   (b)

Fig. 11.22 Spillway segment gates – Arrangement of hydraulic hoist

Fig. 11.23 Hydraulic hoist for the Itumbiara spillway segment gate

In flap gates, the drives are installed as shown in Figure 11.24. In the *a* arrangement, the hoisting mechanism is subjected to tension and may consist of cable, chain or hydraulic hoists. In the *b* case, the hoisting mechanism operates under compression; as a rule, hydraulic cylinders are used in these cases. In the *c* arrangement the hoisting mechanism actuates a lever rigidly connected to the gate torque tube. The lever and the hydraulic cylinder are located in the operating chamber built inside the pier. In any of the above cases, the hydraulic loads on the gate are supported simultaneously by the hoisting mechanism and the bearings.

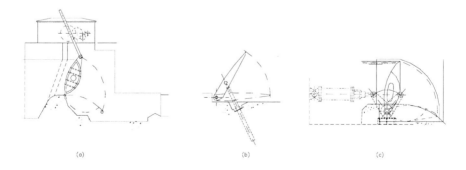

Fig. 11.24 Typical arrangements of flap gate operating mechanism

Fig. 11.25 Hydraulic hoists for the flap gates of the Sihlhofe dam (ZWAG)

11.7 HAND OPERATION

Mechanical and hydraulic hoisting systems should be provided with hand drives to permit the gate closure or opening, in case of power failure. In cable and chain hoists, the supply of power to the electrical circuit must be switched off before the hand drive is engaged to prevent accidents upon power return. Also, the brake should be released only after engaging the hand drive. Hand operation is made through hand wheels or cranks, with a radius of 30 to 40 cm, attached to the speed reducer input shaft. The average speed of crank operation should not exceed 1 m/s. Also, the power developed by the operator should be limited to:
- 100 N-m/s, in continuous operation;
- 150 N-m/s, in operations up to five minutes of duration.

Hydraulic systems are normally provided with a hand pump to supply oil under pressure and allow gate opening in case of power failure.

11.8 DESIGN CRITERIA

11.8.1 LOAD CAPACITY

Hoisting mechanisms are usually designed for operating capacity equal to or over 120 per cent of the maximum load corresponding to the most unfavorable combination of the forces acting on the gate.

11.8.2 OPERATING SPEED

Operating speeds are determined according to the type and purpose of the gate. In spillways, for example, opening and closure speeds of 0.15 to 0.60 m/min for segment gates and of 0.15 to 0.30 m/min for fixed-wheel gates are commonly used.

For intake emergency gates, the following values are usually adopted:
- normal raising and lowering, with the unit at standstill: 1 to 2 m/min;
- emergency closure, against maximum flow through the turbine: 4 to 8 m/min, with damping at the end of the stroke for speeds from 0.3 to 1 m/min.

11.8.3 SAFETY FACTORS

The mechanical components of hoists are dimensioned with a minimum safety factor of five, based on the ultimate strength of the material. For cast iron parts, a minimum safety factor equal to 10 is used.

Steel wire ropes are designed according to the type of loading and the minimum effective breaking strength indicated by the manufacturer. The DIN 19704 standard recommends adoption of the following safety factors:

Load case	Safety factor
Normal	6
Occasional	5
Exceptional	3

Steel wire ropes used in counterweight suspensions are designed with a safety factor of six, for any load case. Roller chains are designed with a minimum safety coefficient of five in relation to the material ultimate strength.

11.9 GATE POSITION MEASUREMENT

The position measurement of vertical lift gates actuated by screw lifts or pedestals can be made by means of a pointer attached to the top of the threaded stem, which displaces vertically along a graduated rule. The measurement is made on a scale of 1:1, that is, the pointer displacement is equal to the gate displacement. The poor visualization of tall rules limits its use to gates with heights up to 3 m.

Fig. 11.26 Position indicator for pedestals (RODNEY HUNT)

For gates with heights exceeding 3 m, the position measurement is usually made by means of a chain or wire cable linked to the gate structure. The linear movement of the chain is transmitted to the receptor by means of sprocket wheels and converted into angular deflection through reduction gears. A rotating pointer on a circular dial displays the gate position. Angular transducers convert angular deflection into an analog signal for remote position indication. The rotating shaft can actuate limit switches.

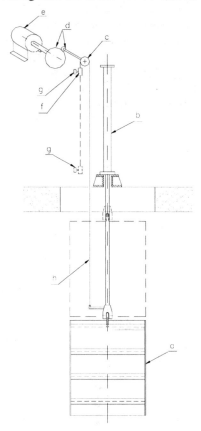

Fig. 11.27 Vertical-lift gate position measurement
(a) gate; (b) hydraulic cylinder; (c) sprocket wheel; (d) reduction gears; (e) angular transducer;
(f) actuator; (g) limit switches; (h) chain

Rexroth Hydraudyne Cylinders B.V. has developed a new technology for position measurement of gates powered by hydraulic cylinders with ceramic piston rod coatings [5]. It consists in a sensor mounted on the cylinder head outside the pressure zone that measures systematic differences in the thickness of the ceramic coating. The pattern of grooves cut in the piston rod base material induces changes in a magnetic field and the additional electronics convert the sensor signal into a logic pulse count. The number of counted pulses represents the position information. In the case of a power loss, the cylinder must be moved to a predefined position, so the position counter can be reset. After reset, the system will keep track of the position. The signal can be read directly into a PC, so that the complete position control of the stroke and synchronization is possible.

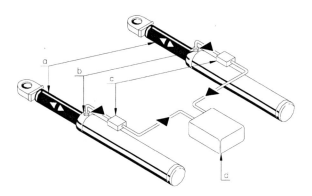

Fig. 11.28 CERAMAX Integrated Measuring System – CIMS (Rexroth Hydraudyne B.V.)
(a) built-in detector; (b) sensor; (c) electronic box; (d) system

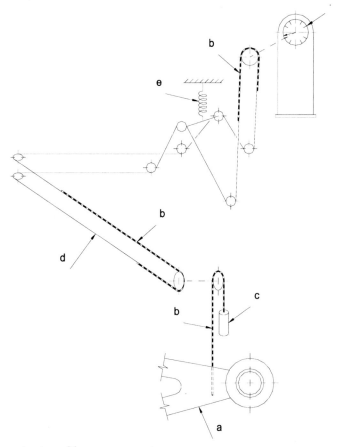

Fig. 11.29 Segment gate position measurement
(a) gate arm; (b) chain; (c) counterweight; (d) wire cable; (e) spring; (f) position indicator

The most common types of position indicators for segment gates are:

a) Mechanical indicator - a chain or a wire cable linked to the gate arms actuates a sprocket wheel that operates the position indicator located on the operating deck, by converting the linear displacement of the chain into rotary motion. A sliding or a rotating pointer indicates the gate position. A counterweight is used to keep the chain stretched by pulling. The accuracy of the mechanical reading is not high due to the possibility of the chain stretching with use or by variation of the temperature.

b) Linear transducer - a sliding shoe attached to the gate arm converts the angular movement of the gate into a linear displacement of a shaft connected to a linear position transducer. The shoe is vertically guided in the interior of a structural metal case installed on the pier face or on the trunnion support. The linear transducer offers high accuracy and provides an electrical analog signal for telemetry. The gate-opening stroke is directly proportional to the vertical displacement of the shoe.

c) Angular detector - it measures the angle of rotation of the gate and comprises a free-rotating eccentric mass installed in a closed case rigidly attached to the gate moving part. The mass remains in the vertical position whatever the gate position, and actuates a potentiometer through reducing gears. The angular deflection between the mass and its encasement is converted into an analog signal for monitoring, recording and remote position indication. The angular detector is used to measure the angle of rotation of flap and radial gates, valves and movable bridges.

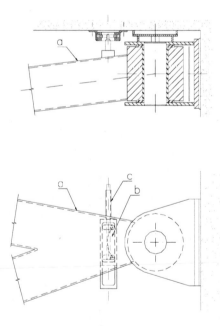

Fig. 11.30 Linear transducer for gate position measurement
(a) gate arm; (b) sliding shoe; (c) linear transducer

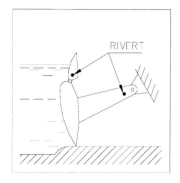 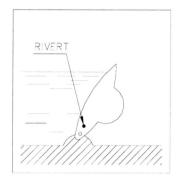

Fig. 11.31 RIVERT-type angular detector (Rittmeyer)

REFERENCES

1.Faires, Virgil M.: *Design of Machine Elements*, The MacMillan Co., New York (1963).
2.CIMAF: *Wire Ropes General Catalogue* (in Portuguese).
3.Rudenko, N.: *Materials Handling Equipment*, Peace Publishers, Moscow.
4.Korotikov, Boris: *Modern Design of Intake Gate Hydraulic Hoists* (in Portuguese), Rio de Janeiro (Feb. 1981). Not published.
5.Perez, P.M. and van de Ven, R.: Ceramic Coating for Intake Gate Cylinders, *International Water Power and Dam Construction* (Dec. 1994).

Chapter 12

Materials

12.1 INTRODUCTION

In the manufacture of gates and appurtenances a great variety of materials, such as rolled, stainless, cast and forged steels, cast iron, bronze and its alloys, natural and synthetic rubber for seals and, in some cases, timber, is used.

The wide range of types and the amount of materials available can satisfy all the needs and requirements of the gate designers, thus leading to an adequate and economic selection of the material. This selection must comply with sound criteria, in order to warrant reliable operation and long life for the equipment. Therefore, the designer should take into account not only the cost and availability of the material in the market but also its main characteristics such as the chemical composition, physical and mechanical properties (yield and tensile strength, elongation, toughness, hardness, weldability, machinability, heat treatment capability) and the corrosion resistance.

12.2 HEAT TREATMENT

The heat treatment of steel permits the alteration of its basic features (mechanical resistance, hardness, granular composition, ductility, toughness, wear resistance) as well as the removal of internal residual stresses.

Annealing consists of heating of the steel to an appropriate temperature (as a function of the carbon content), holding it at that temperature for a suitable time and then slowly cooling it in the furnace. Annealing may produce a required particular microstructure and also restores the original properties to the steel modified by heat treatment.

Normalizing produces a uniform structure and consists in the heating with subsequent air cooling, at room temperature. Hardness and mechanical properties are higher than in the case of annealing.

Quenching is the heating of the material above the transformation range, followed by sudden cooling in oil or water, for the purpose of increasing hardness and mechanical resistance. It reduces the toughness and ductility, and may create internal stresses.

Tempering is the re-heating of hardened or normalized steel at a temperature below the transformation range, followed by slow cooling. In general, hardened steels are tempered in order to relieve internal stresses, to restore part of the ductility, and to improve toughness.

Carburizing is a process of surface hardening and consists of adding carbon to the surface of steel by exposing it to carbon rich substances at a suitable temperature for a long period of time. Quenching usually follows carburizing.

12.3 ROLLED STEELS

Rolled steel plates and shapes comprise the greatest portion of weight of the material used in the gate construction and are used for manufacture of elements with structural function such as skin plate, girders, lateral guides, wheel track supports and so on. They may be classified in two groups:

a) structural quality - these are steels of medium resistance and good weldability. The following types of steel fit into this group: ASTM A36 and A283 Grade D; DIN 17100 R St 37.2 and R St 42.2. The most commonly used is the ASTM A36, with minimum yield strength of 245 MPa and a tensile strength varying from 392 to 539 MPa. This steel does not require heat treatment, except stress relieving of welded parts to be machined;

b) high-strength weldable structural quality - these are steels with greater mechanical resistance than the usual structural quality steels, in addition to higher toughness. Owing to their low carbon content (about 0.2 per cent), these steels also have good weldability. The use of these steels, when compared to those of common grades, provides reduction of thickness and, consequently, of weight. In this group are the steels DIN 17100 RR St 52.3, ASTM A441 and NTU-SAR-50 and 55.

Certificates, called mill sheets, issued by the steelworks usually state the chemical composition and the mechanical properties of steels.

12.4 STEELS FOR MACHINE ELEMENTS

These are carbon and alloy steels used in manufacturing parts with mechanical functions, like gears, pinions, shafts, wheels and pins. They can be rolled, cast or forged, and also heat-treated. Rolled carbon steels of the series AISI 1020, 1030, 1040 and 1045 are commonly employed with no heat treatment.

In the manufacture of parts subjected to shocks and severe wear, it is usual to employ low carbon steels (generally up to 0.25 per cent), carburized, quenched and tempered between or after the machining operations. This heat treatment provides a hard surface allied to a tough core. As an example, the carburizing of parts manufactured of nickel chromium molybdenum steel AISI 4320 leads to a core tensile strength from 1080 to 1275 MPa.

Steels with a carbon content above 0.25 per cent are frequently heat treated for improvement of their mechanical properties, and are recommended for the manufacture of mechanical parts that require a good combination of strength and toughness. The heat treatment usually comprises quenching followed by tempering. Among these are the manganese steel AISI 1340, the chrome molybdenum AISI 4140 and the nickel chromium molybdenum AISI 4340.

12.5 STAINLESS STEELS

Due to their good corrosion resistance imparted by the presence of a high chromium content in their constitution (more than 10 per cent), these steels are used in the manufacture of seal seats, servomotors stems, wheel tracks and parts needing periodical removal such as bolts, pins and shafts.

The austenitic steels (AISI 304, 304L and 316, for example) contain both chromium and nickel (total content not less than 23 per cent) and are used where good corrosion resistance and medium structural resistance are required (seal seats, seal fastening bolts and so on); the AISI 304 does not resist salt water. These steels can be easily welded; however, acetylene welds should be avoided. Parts manufactured with 304 steel should preferably be annealed after welding. The 304L do not require any heat treatment after welding, except in parts that will be exposed to the attack of aggressive agents, which is not the case of hydromechanical equipment. The 316 type is the austenitic steel of greatest resistance to corrosion. It resists acids and is recommended for applications in contact with salt water.

The martensitic steels (AISI 410 and 416) contain only chromium as the primary alloy and are recommended for the manufacture of parts in which high physical properties are more important than the corrosion resistance obtained with the austenitic alloys, as for example, hydraulic cylinder stems. Both types present their best corrosion resistance features in the quenched state and with the surface finely polished. The 416 type has about the same chemical composition, the same mechanical properties and the same corrosion resistance as the 410, from which it is distinguished by the addition of sulphur, which gives it excellent machinability. Welding of these steels requires preheating and heat treatment. The parts must be preheated from 200 to 300 degrees C and, during welding, the temperature must not drop to less than 200 degrees C. Immediately after welding, the part should be annealed to about 700 degrees C, with slow cooling, for removal of residual stresses. Type 410 has the lowest cost of the stainless steels.

12.6 CAST STEELS

These are generally used in the manufacture of complex geometry or bulky parts, or also high loaded parts, such as wheels, trunnion hubs for segment gates, bearing supports, heads of hydraulic cylinders, hook eyes of cylinder stems and roller supports of caterpillar gates.

Cast parts of average and low strength may be manufactured according to ASTM A27, while the ASTM A148 is used in parts which need greater strength. As a rule, cast parts are supplied heat-treated by the foundry, annealed or normalized.

12.7 FORGED STEELS

These are recommended for special cases of heavy loaded parts such as wheels of fixed-wheel gates, main pins of segment gates and lifting eyes. Under this heading are the ASTM A105, A236 and A668 steels.

12.8 GRAY CAST IRONS

Cast iron is an alloy of iron and carbon (1.7 to 5 per cent of C) formed through casting of pig iron with scrap. Gray iron is characterized by the presence of graphite (pure carbon) while in the white cast iron the carbon takes the form of cementite (Fe_3C). Gray cast iron has excellent wearing properties, is easily machined and has good corrosion resistance. Due to the high content of carbon in its composition, gray iron is not malleable.

It is used in the manufacture of bearing supports, cable hoist drums, gears, sheaves, pedestals, small guide wheels and in gate structure and embedded parts of small slide gates (up to 3 m x 3 m) and water heads up to 20 m. Its use is not recommended in parts subjected to heavy shocks or waterhammer. Cast parts are usually specified in accordance to the ASTM A48 and A126 standards.

12.9 BRONZES

These are copper and tin alloys that are easily machined, have good corrosion resistance and anti-friction characteristics, being chiefly suitable for use in movable parts subject to compression.

The main alloys used in gates are:
a) High-lead tin bronzes - recommended for the manufacture of bushings, guide shoes and slide gate seals. The most common specification of this alloy is the ASTM B-584-937, which contains 80 per cent copper, 10 per cent lead and 10 per cent tin;
b) Manganese bronzes - these alloys have high mechanical and corrosion resistance, being often used in worm gears (ASTM B-584-862) and self-lubricating bushings (ASTM B-22-863);
c) Aluminum bronzes (ASTM B-148) - these alloys combine high mechanical strength with high ductility and low coefficient of friction. Their main use is in the manufacture of stem lift nuts.

12.10 BOLTS

Bolts used in the manufacture of gates can be broadly classified in four groups:
a) Common bolts - these are the bolts subjected to low or average loads, used in joints intended or not for field welding, or that do not require periodical removal. In these cases carbon steels of the series AISI 1020, 1030 and 1035, and also the ASTM A307 are used. Bolts of dismountable joints should be made of stainless steel;
b) Heavy-duty bolts - are those used in high loaded connections (heads of hydraulic cylinders, for example). Among others, the AISI 4130 and 4820 steels can be used;
c) High strength connection bolts - are used in structural parts to be connected by friction to the contact surfaces through an initial tightening load. These bolts are manufactured according to the ASTM A325 standard;
d) Seal bolts - due to the need of periodical removal for seal adjustment or replacement; they should have anticorrosive characteristics and may be made of stainless steel (AISI 304 or ASTM A-193-B6) or brass.

Table 12.1 Rolled Steels for Structural Elements

Specification			Thickness range (mm)	Chemical composition (% max.)					Mechanical properties (MPa)		Typical applications
				C	Si	Mn	P	S	YS	TS	
ASTM	A 36		from 5 to 152.4	0.25/0.29	0.15/0.40	0.85/1.20	0.040	0.050	250	400	Low and average loaded structures
	A 283	Gr. D	from 5 to 152.4	-	-	-	0.040	0.050	230	415	
DIN	17100	RSt-37.2	from 5 to 75	0.17	-	-	0.050	0.050	235	340	
		RSt-44.2	from 5 to 75	0.22	-	-	0.050	0.050	275	410	
			from 5 to 19.1						345	480	
ASTM	A 441		from 19.1 to 38.1	0.22	0.40	0.85/1.25	0.040	0.050	315	460	Average and heavily loaded structures
			from 38.1 to 50.8						290	435	
DIN	17100	RRSt-52.3	from 5 to 150	0.22	0.55	1.60	0.040	0.045	325	490	
NTU	SAR-50	A	from 5 to 75	0.20	0.55	1.2/1.5	0.030	0.030	330	500	
	SAR-50	B	from 5 to 75	0.20	0.55	1.2/1.5	0.030	0.030	330	500	
	SAR-55		from 5 to 32	0.18	0.55	1.2/1.5	0.030	0.030	360	550	

Legend: NTU - USIMINAS Technical standards
YS - Yield strength; TS – Tensile strength

Table 12.2 Steels for Mechanical Parts

ASTM	Condition (a)	Chemical Composition (%) (b)							Mechanical Properties			Typical applications
		C	Mn	P max.	S max.	Ni	Cr	Mo	YS (MPa)	TS (MPa)	Brinell Hardness (c)	
1020	As rolled	1.18/0.23	0.3/0.6	0.04	0.05	-	-	-	331	448	143	Anchors, bolts and ...
1030	As rolled	0.28/0.34	0.6/0.9	0.04	0.05	-	-	-	345	552	179	Pins, shafts, wheels, small gears and pinions
1040	As rolled	0.37/0.44	0.6/0.9	0.04	0.05	-	-	-	414	621	201	
1045	As rolled	0.43/0.50	0.6/0.9	0.04	0.05	-	-	-	496	662	215	
4320 (d)	C 930°C DOQ	0.17/0.22	0.45/0.65	0.04	0.04	1.65/2.0	0.4/0.6	0.2/0.3	669	1047	302	Heavily loaded
1340 (d)	OQ 830°C T 650°C	0.38/0.43	1.6/1.9	0.04	0.04	-	-	-	748	817	241	mechanical parts and/or subjected to
4140 (d)	OQ 840°C T 650°C	0.38/0.43	0.75/1.0	0.04	0.04	-	0.8/1.1	0.15/0.25	888	941	277	impact and heavy
4340 (d)	OQ 800°C T 650°C	0.38/0.43	0.6/0.8	0.04	0.04	1.65/2.0	0.7/0.9	0.2/0.3	935	1000	285	surface wear

Legend:
(a) C – Carburized; OQ – Oil quenched; DOQ – Double oil quenched; T – Tempered;
(b) Before heat treatment
(c) Core hardness after hear treatment
(d) ½" diameter, 2" long
YS – Yield strength; TS – Tensile strength

Table 12.3 Stainless Steels

Specification		Chemical composition (%)							Mechanical properties (a) (MPa)		Typical applications
		C max.	Mn max.	Si max.	P max.	S	Cr	Ni	YS	TS	
	304	0.08	2.0	1.0	0.045	0.03	18/20	8/10.5	245	590	Welded parts, seal seats, pins, shafts and bolts
	304L (b)	0.03	2.0	1.0	0.045	0.03	18/20	8/12	245	590	
AISI	316	0.08	2.0	1.0	0.045	1.0	16/18	10/14	245	590	Parts in contact with salt water
	410	0.15	1.0	1.0	0.040	0.03	11.5/13.5	-	295	490	Cylinder rods and wheel tracks
	416	0.15	1.25	1.0	0.06	0.15	12/14	-	295	490	

Legend:
(a) Annealed, Brinell hardness of approx. 160;
(b) Molybdenum-content: from 2.0 to 3.0 per cent;
YS Yield strength;
TS Tensile strength

Table 12.4 Steel Casting and Forgings

Type	Specification		Condition	Chemical composition (% max.)					Mechanical properties (MPa)		Typical applications
				C	Mn	Si	S	P	YS	TS	
SC	A 27	Gr.60-30	A	0.30	0.60	0.80	0.06	0.05	207	414	Wheels, trunnion hubs for segment gates, bearing supports, heads of hydraulic cylinders, hook eyes of cylinder rods
		Gr.65-35	N	0.30	0.70	0.80	0.06	0.05	241	448	
		Gr.70-40	N	0.25	1.20	0.80	0.06	0.05	276	483	
	A 148	Gr.90-60	N and T	(a)	(a)	(a)	0.06	0.05	414	621	
		Gr.105-85	Q and T	(a)	(a)	(a)	0.06	0.05	586	724	
ASTM	A 105	Gr. I	A or N	0.35	0.90	0.35	0.05	0.05	207	414	
		Gr. II	A or N	0.35	0.90	0.35	0.05	0.05	248	483	
	A 236	Gr. E	N and T	0.4/0.55	0.6/0.9	0.15(b)	0.05	0.045	296	572	Wheels, pins for segment gates, hooks, lifting eyes, hydraulic cylinder heads
		Gr. F	Double N and T	0.45/0.59	0.6/0.9	0.15(b)	0.05	0.045	317	580	
SF		Gr. G	Q and T	-	0.6/0.9	0.15(b)	0.05	0.045	344	587	
		Gr. H	N, Q, T	-	0.6/0.9	0.15(b)	0.05	0.045	414	690	
	A 668	Gr. D	N, A or N and T	(a)	1.10	(a)	0.05	0.05	260	515	
		Gr. H	N and T	(a)	-	(a)	0.04	0.04	400	620	
		Gr. L	N, Q, T	(a)	-	(a)	0.04	0.04	585	760	

Legend:
(a) At discretion of the supplier
(b) Minimum content
SC – Steel castings; SF – Steel Forgings; ; T – Tempered; TS – Tensile strength; YS – Yield strength

Table 12.5 Gray Cast Irons

Specification		Chemical composition, % (a)		Tensile strength (MPa)	Typical applications
		P max.	S max.		
A 48	30			207	Bearings, supports, cable hoist drums, floor stands, gate leaf and seat facings of small slide gates, gears and sheaves
	40			276	
	50			345	
ASTM	Class A	0.75	0.15	145	
A 126	Class B	0.75	0.15	214	
	Class C	0.75	0.15	283	

Legend:
(a) carbon-content: from 1.7 to 5 per cent.

Table 12.6 Bronzes and Brass

Type	Specification		Nominal chemical composition (%)							Mechanical properties			Typical applications
			Cu	Sn	Pb	Zn	Fe	Al	Mn	YS (MPa)	TS (MPa)	Brinell hardness (500 kgf)	
		937	80	10	10					83	207	65	
High-lead tin bronzes	ASTM B 584	932	83	7	7	3				97	207	62	Bushings, guide shoes and seals for gates
		935	85	5	9	1				83	193	52	
Manganese bronzes	ASTM B 584	862	63	0.2 max.	0.2 max.	27	3	4	3	310	621	180	Worm gears
	ASTM B 22	863	60/68	0.2 max.	0.2 max.	rest	2/4		3/7.5	414	758	223	Self-lubricating bushings
Aluminum bronzes	ASTM B 148	954	85				4	11		207	517	150 (b)	
		955 (a)	81				4	11		276	621	190 (b)	Stem lift nuts
Naval brass	ASTM B 21	464	59/62	0.5/1	0.2 max.	rest	0.1 max.			138	345		Seal bolts

Legend:
(a) Nickel content = 4 per cent;
(b) Load of 3000 kgf
'rest' = remainder; TS – Tensile strength; YS – Yield strength

Chapter 13

Gate Seals

13.1 INTRODUCTION

Seals are the elements placed between the gate leaf and the embedded frame in order to prevent leakage. Wood, metal and rubber have been used as base materials for seals.

13.2 WOOD SEALS

Wood was certainly the first material employed as seals in the early projects. Nowadays, its use is restricted to small gates or to specific applications such as the lower lip seal of rolling gates or the vertical seal between the two miter gate leaves.

13.3 METALLIC SEALS

When used in slide gates, metal seals are designed to provide water tightness and, at the same time, to transfer to the embedded parts the hydraulic load acting on the gate leaf. These seals are made of rectangular bars of bronze, brass or stainless steel bolted to the skin plate. Seal and seat surfaces are both machined after all shop welding has been completed.

13.4 RUBBER SEALS

In modern practice, gate seals are mostly made of rubber, due to its durability, great flexibility and satisfactory strength. The rubber seal may be fabricated in any desired shape, thus allowing the designer to choose the best shape for every application. In order to reduce costs, however, the seal section should be one that can be made by most manufacturers from their existing molds.

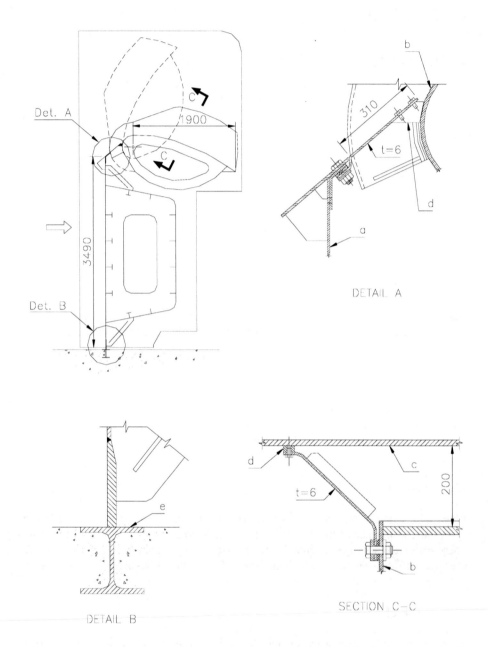

Fig. 13.1 Fixed-wheel gate with flap at the Sambre river lock, with metallic seals
(a) fixed-wheel gate skin plate; (b) flap gate skin plate; (c) side apron; (d) metallic seal (white metal);
(e) sill beam

Figure 13.2 shows the four basic profiles of rubber seals: J-type (or music note), J-type with double stem, angle and rectangular shapes. There is no rigid rule governing the type of seal profile that should be used for each type of gate.

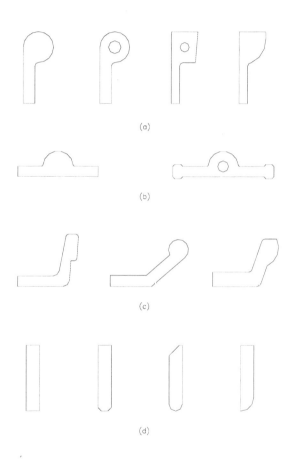

Fig. 13.2 Rubber seal types (RUBBERART)
(a) J-type or music note; (b) J-type double stem; (c) angle-shaped; (d) rectangular

Generally speaking, the most common applications of rubber seals are:
a) Single stem J-type (or music note) - top and side seals of low-head gates;
b) Double stem J-type - top and side seals of high-head gates;
c) L-type (or angle-shaped) - side seals of weir gates;
d) Rectangular - bottom seals.

J-type seals are available with both hollow and solid bulbs. Hollow bulb seals provide a greater contact area with seats, thus aiding water tightness in low-head gates, since the hydrostatic load on the seals is small, in this case. It is observed that the compression of hollow bulbs over long periods of time may result in difficulty of restoring the original seal shape when the pressure is removed. In solid bulb seals, however, this effect is negligible. In general, hollow bulb seals are more costly than solid ones. At splices of hollow bulb seals, soft rubber plugs are introduced and cemented in the hole. These plugs allow the manufacturer to make correct cuts in adjusting the seals to length and facilitate cementing and vulcanizing of splices. The plugs are inserted into holes before final trimming of the seal.

Rubber seals are commonly attached to the gate by means of clamp bars and bolts through holes drilled in the seal stem. Bolt spacing is usually taken about 100 mm.

Bolt materials with non-corrosive characteristics facilitate bolt removing even after prolonged periods of gate immersion. Stainless steel, brass and bronze are some options for bolt materials.

Since drilled bolt holes in the skin plate are potential leakage points, the use of sealing washers between the nut and the skin plate is advisable. These washers are generally made of lead or rubber.

Rubber gate seals are frequently supplied without bolt holes, which are then drilled through the gate skin plate, seal base and clamp bars with the help of jig plates, after completion of the manufacture.

In some recent projects of weir segment gates, bolt holes are not drilled through the seal stems. L-shaped side seals are then held by the compression developed between a clamping bar and the seal base, thus assuring an easier seal adjustment. When floating debris are present, side seals of weir segment gates are protected by metallic angles, as shown in Fig. 13.3.

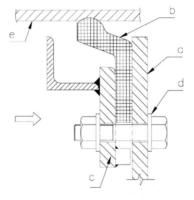

Fig. 13.3 Segment gate – Side seal and protector angle (DEDINI)
(a) skin plate; (b) seal; (c) clamp bar; (d) washer; (e) seal support

When closing a sluice gate under unbalanced head, the seal may be jammed into the clearance space between the clamp bar and the seal plate due to the venturi action, thus causing the seal to be damaged. The tendency of the seal to enter the gap between the seal plate and the clamp bar is reduced by using a shaped clamp bar, as shown in Fig. 13.4a, or alternatively, with the use of double stem J-seals (Fig. 13.4b). Two clamp bars bolted to the gate skin plate make the seal attachment. The seal is actuated by water pressure from the reservoir and admitted to the pressure chamber through holes in the seal base.

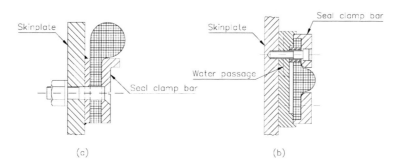

Fig. 13.4 Seals for high-pressure gates

13.5 MATERIAL FOR RUBBER SEALS

Gate rubber seals are molded from natural or synthetic rubber compounds (styrene-butadiene or Neoprene, for example). A rubber compound consists of many ingredients. Its most important component, the base elastomer, has certain inherent advantages and limitations, but carbon black, zinc oxide, antioxidants, vulcanizing agents, accelerators, and others special purpose additives also affect a compound's final properties. Such products are added to the base elastomer in accordance to the manufacturer experience so as to provide suitable characteristic for every application. In general, the physical characteristics of both natural and synthetic rubbers are equivalent.

The properties of the natural and the synthetic rubbers are compared in the Table 13.1.

Table 13.1 Comparative Properties of Natural and Synthetic Rubbers

Properties	Natural rubber	Neoprene	Styrene-Butadiene
Tension strength, MPa	10.3 to 27.5	10.3 to 27.5	10.3 to 27.5
Hardness range, Shore A	30 to 90	40 to 95	40 to 100
Specific gravity of base elastomer	0.93	1.23	0.94
Abrasion resistance	Excellent	Excellent	Excellent
Oil resistance	Poor	Good	None
Resistance to weathering	Poor	Excellent	Good
Low temperature resistance	Excellent	Good	Good

Source: G.E. Technical Data Book S-1A - Silicones

13.6 CLAD SEALS

Metal or plastic film rubber seal cladding are used to reduce the friction coefficients or to prevent distortion of the seal bulb into the clearance between the gate and the seal seats. Brass or stainless steel sheaths or fluorocarbon film are introduced into the mould along with the raw non-vulcanized rubber compound, and then molded and vulcanized simultaneously with the rubber. A strong vulcanized bond between the metal or film and the rubber results during this process. Metal sheath thickness is limited to 1.5 mm, since thicker metal adds more rigidity to the seal. Thickness of fluorocarbon film varies from 0.8 to 1.5 mm.

Manufacturing costs of metal clad seals are relatively high. It is observed that the metal on the sealing face does not prolong the life of the seal.

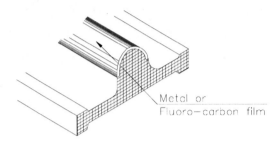

Fig. 13.5 Clad seal

13.7 RUBBER SEAL HARDNESS

Rubber seals are supplied with hardness varying from 45 to 80 Shore A. As a rule, non-sliding seals call for lower hardness, whereas sliding seals require higher hardness. Friction coefficient between seals and seats depends on the rubber hardness. The German standard DIN 19704 reports the result of tests carried out on friction coefficients for dry friction, slightly rusted surfaces and rubber seals of various hardness, as follows:

Shore A hardness	Friction coefficient
70	0.80
55	0.90

Conservative values of friction coefficients of rubber on steel, and fluorocarbon film on steel are given in Chapter 9.

13.8 RUBBER SPECIFICATIONS

The following is a typical specification for gate rubber seals:
- Tensile strength   20.6 MPa   (min.)
- Elongation at break   450 %   (min.)
- Durometer hardness Shore A   55 to 65
- Compression set, % of original deflection   30 %   (min.)
- Tensile strength after accelerated aging,
  percentage of initial tensile strength   80 %   (min.)
- Water absorption by weight   5 %   (max.)

The fluorocarbon film specifications should have the following physical properties:
- Tensile strength   13.7 MPa   (min.)
- Elongation   250 %   (min.)

13.9 SEAL LEAKAGE

As a rule, the maximum seal leakage of a gate shall not exceed 0.1 liter/second per meter of seal length. In a 5 m by 8 m gate, for instance, the maximum acceptable leakage would be:
- total length of seals: 2(5+8) = 26 m
- allowable leakage: 0.1 x 26 = 2.6 liters/second.

At the Inguri low level outlets, a leakage of 5 to 10 liters/second was observed at the 9 m by 10 m service gate, located 181 m below the maximum reservoir level. This corresponds to a rate of leakage from 0.13 to 0.26 liters/second/meter of seal [1].

13.10 MANUFACTURE AND ASSEMBLY OF SEALS

In most cases, rubber seals are molded and splices are made in a vulcanizing press, preferably at the factory, under controlled conditions. Wherever practical, splices should be on a 45-degrees bevel as related to the thickness (and not to the width) of the seal. Splices should be located midway between bolt holes.

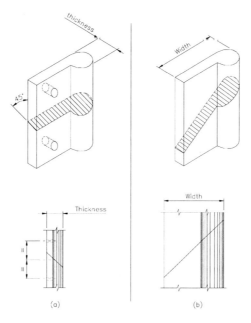

Fig. 13.6 Seal splices
(a) correct; (b) to avoid

Non-vulcanized field splices are rare and should be cut off square and never on a bevel. In these splices, seals should be cut slightly longer than required so as to permit fitting without buckling or misalignment. As an additional precaution against leakage, rubber cement may be applied during the final attachment of the seals to the gate.

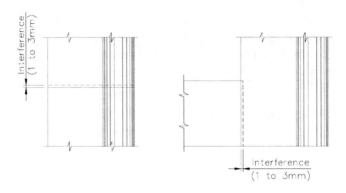

Fig. 13.7 Non-vulcanized field splices

Seals are preferably fastened to finished areas in such a manner as to ensure effective sealing over their full length. All steel members that are in contact with, or upon movement, may come in contact with the seals should have their edges rounded. The seal plates for the side seals are dimensioned such that seals remain in effective contact over the range of sideways movement allowed by the lateral clearance. The assembly of seals of downstream sealing gates should accommodate deflections without damage, when under load. On the other hand, special care should be given to the top seal design of upstream sealing gates, since the deflection towards downstream of the top portion of the gate leaf may create a gap between the top seal and the upstream lintel.

Wherever possible, J-type seals should be mounted to utilize stem deflection rather than bulb compression, since this may cause crushing or shearing of the bulb seal. Preset of the bulb should be held to the minimum necessary to accomplish sealing. A preset of 5mm or less is adequate.

In the following figures, some examples of seal application for various types of gates are shown.

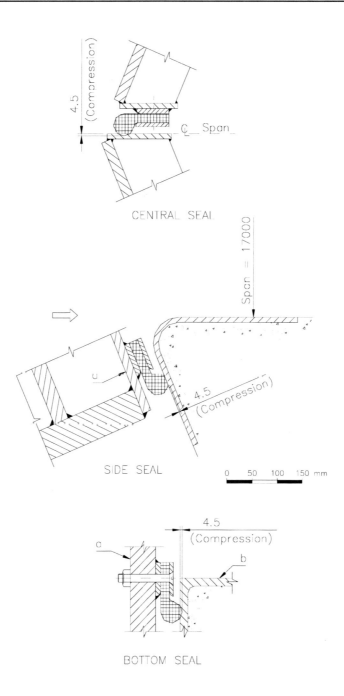

Fig. 13.8 Miter gate at Sobradinho Lock (ALSTOM), 17 m wide by 38 m high
(a) skin plate; (b) sill beam

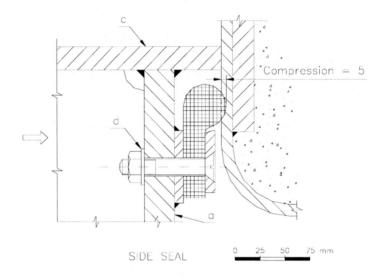

Compression = 5

SIDE SEAL

0   25   50   75 mm

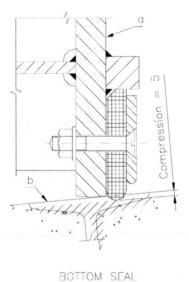

Compression = 5

BOTTOM SEAL

Fig. 13.9 Spillway stoplogs at Porto Colombia Power Plant (Ishibras)
(a) skin plate; (b) sill beam; (c) vertical end girder; (d) washer

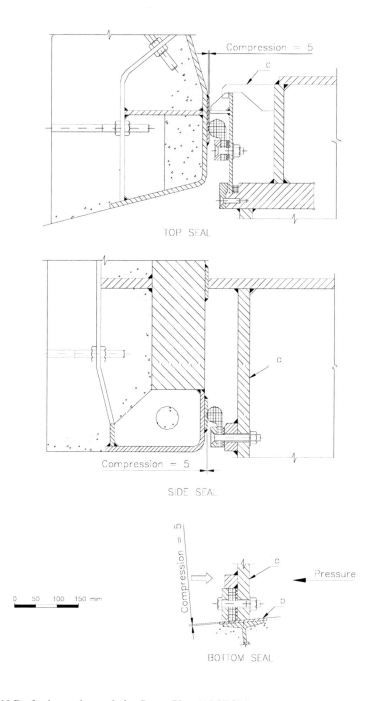

Fig. 13.10 Draft tube stoplogs at Itaipu Power Plant (ALSTOM)
(a) skin plate; (b) sill beam; (c) limiter

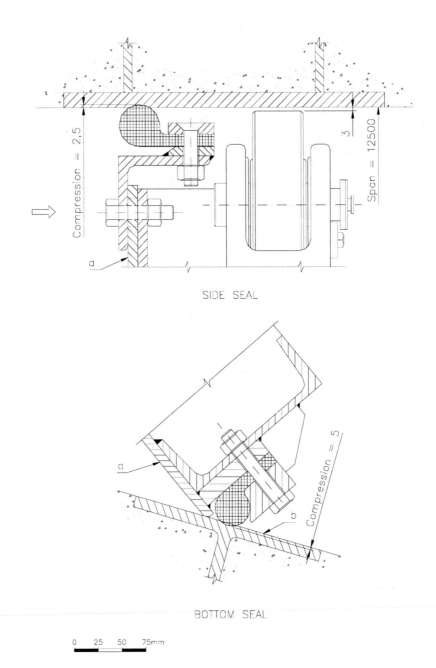

SIDE SEAL

BOTTOM SEAL

0   25   50   75mm

Fig. 13.11 Spillway segment gate at Euclides da Cunha Power Plant (ALSTOM)
(a) skin plate; (b) sill beam

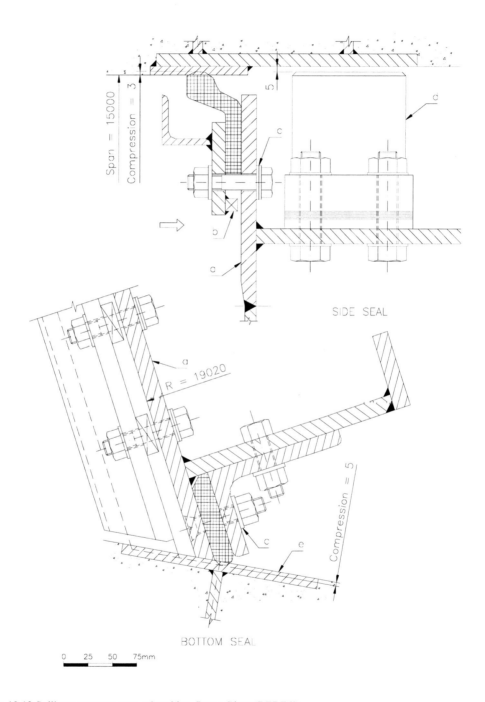

Fig. 13.12 Spillway segment gate at Itumbiara Power Plant (DEDINI)
(a) skin plate; (b) stop plate; (c) Neoprene washer; (d) lateral shoe; (e) sill beam

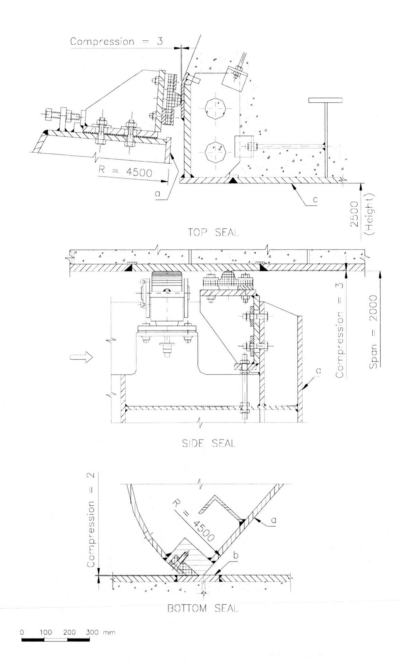

Fig. 13.13 Reverse segment gate for the aqueducts of the Nova Avanhandava lock (ALSTOM)
(a) skin plate; (b) sill beam; (c) steel liner

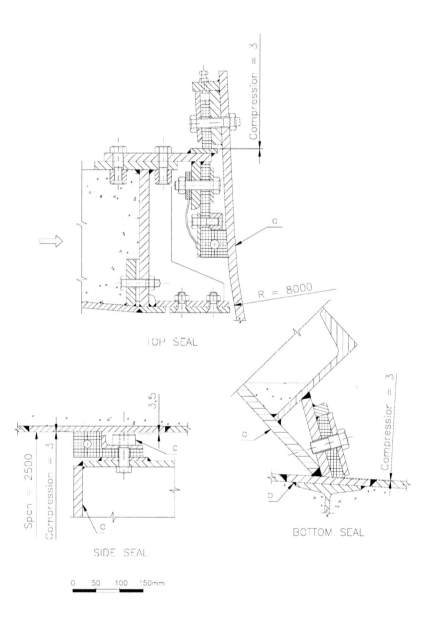

Fig. 13.14 Bottom outlet segment gate at the Parigot de Souza Power Plant (Ishibras)
(a) skin plate; (b) sill beam; (c) bronze shoe

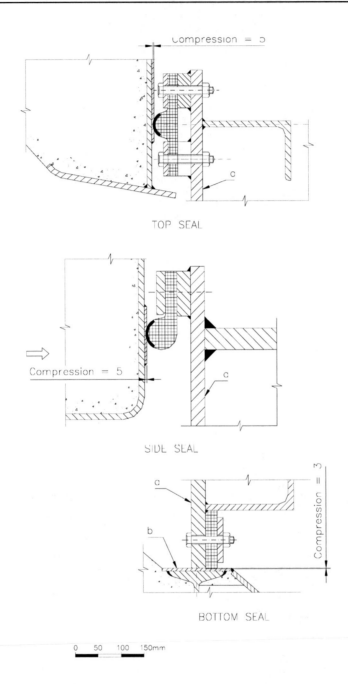

TOP SEAL

SIDE SEAL

BOTTOM SEAL

0   50   100   150mm

Fig. 13.15 Intake fixed-wheel gate with upstream seals at the Nova Avanhandava Power Plant (ALSTOM)
(a) skin plate; (b) sill beam

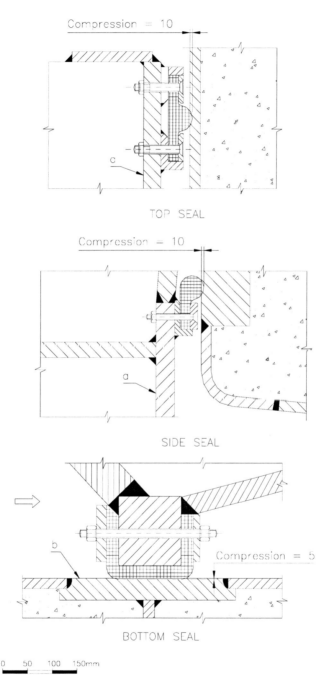

Fig. 13.16 Fixed-wheel gate with downstream seals for the Itaipu diversion sluiceway (ALSTOM/Bardella)
(a) skin plate; (b) sill beam

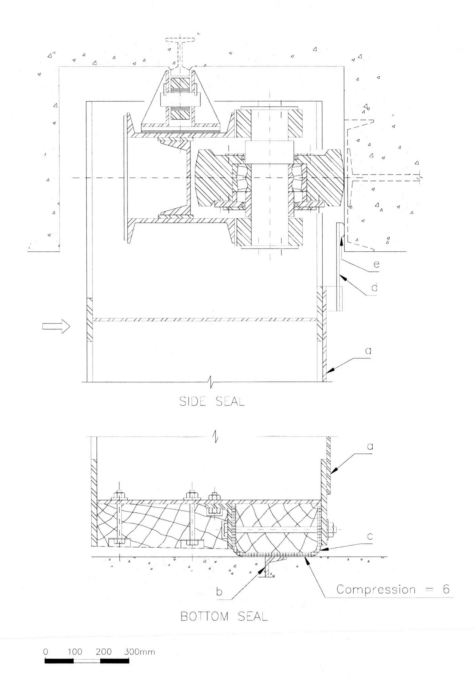

SIDE SEAL

BOTTOM SEAL

Compression = 6

0   100   200   300mm

Fig. 13.17 Intake fixed-wheel gate at the Mascarenhas de Moraes Power Plant (Ishibras)
(a) skin plate; (b) sill beam; (c) rubber bottom seal; (d) stainless steel plate; (e) brass side seal

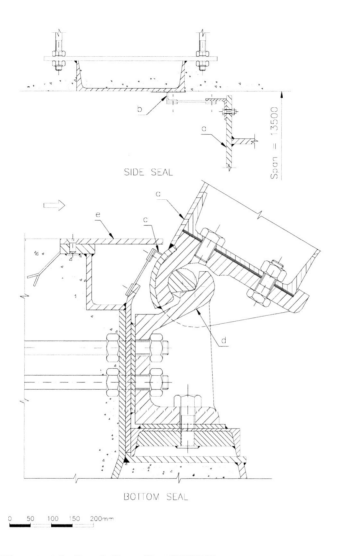

SIDE SEAL

BOTTOM SEAL

0   50   100   150   200mm

Fig. 13.18 Flap gate at the Serraria Power Plant (DEDINI)
(a) skin plate; (b) side seal; (c) bottom seal; (d) hinge support; (e) sill beam

REFERENCE

1. Farmakovsky, S.A. and Kaplan, I.H.: Low-down on Inguri, *International Water Power & Dam Construction* (Sept. 1994).

Chapter 14

Manufacture, Transportation and Erection

14.1 MANUFACTURE

14.1.1 MANUFACTURING STEPS

The manufacture of gates consists basically of steelwork and machining services. Before starting manufacture, production flowcharts for the various components or equipment are prepared, with the following main steps:

. full-size layout drawings;
. storing of raw material;
. marking;
. cutting;
. curving;
. welding;
. finishing;
. pre-assembly;
. machining;
. mechanical fit-up;
. anticorrosive protection;
. inspection.

Manufacture is always based on drawings approved by the purchaser.

14.1.2 FULL-SIZE LAYOUT DRAWINGS

Consists of the preliminary works for cutting and curving of plates, rolled sections, pipes and so on, and comprises the full-scale drawing of pieces, the making of templates and preparation of the cutting plan (where one tries to obtain the best use of the raw material, particularly in the case of repeated parts) and the issuing of bill of materials.

14.1.3 STORING OF RAW MATERIAL

In large manufacturers, a list of material in stock is periodically sent to the design department, through which the designer makes his choice, observing the limitations and requirements imposed by the specifications. In the warehouse, materials are duly sorted and designated for the current work. Materials ordered by the designers and not in stock are then acquired in the market.

14.1.4 MARKING

The marking starts when the material, the cutting plan and the templates are available. All necessary information, such as cutting dimensions, edge preparation details and piece marks, are then transferred to the part to be worked on. Marking is usually performed with white paint. In certain cases, metal tracers and punchers are used.

14.1.5 CUTTING

This phase consists basically in cutting the parts according to the information written or traced in the material. The usual cutting processes are:
a) oxygen-cutting and plasma arc (for non-ferrous and stainless steel):
  - automatic: parallel cutting bench;
  - pantographic bench (numerical control, optical, magnetic);
  - semiautomatic: moving track devices ('turtles') for straight and curved cuts;
  - manual: torch;

b) machine cutting:
  - shearing;
  - saw;
  - disc.

14.1.6 CURVING

After the plates and rolled sections are cut to size, some parts must to be curved with the help of hydraulic presses, plate rollers or torches.

In general, the spherical and conical shapes are made in the hydraulic presses, as well as the curving of the initial and final edges of plates to be rolled. Plate rollers are used to make the cylindrical curves of plates and rolled sections, and the conical curving. Reverse shapes are obtained with the use of torches. Curving templates are previously prepared for the three processes.

14.1.7 STRUCTURE WELDING

Qualification procedures for welding processes and welders are developed before starting fabrication. It is an accepted practice that all welders to be assigned to the work must be duly qualified according to the welding process to be applied.

Fabrication of the structure starts with the separate welding of the sub-assemblies. Skin plates of large dimensions and webs formed by plates of different thicknesses are butt-welded. The flanges of T-beams are welded to the respective webs. After welding, the sub-assemblies are inspected and straightened, if necessary. Then comes the fabrication of the final structure, which consists of the positioning and welding of the sub-assemblies. If several identical parts are to be made, it is often useful to prepare jigs or forms on which the sub-assemblies are set up. This procedure is taken, e.g., for fabrication of vertical guides, frames of flat gates and skin plate framing of spillway gates (see Figure 14.1).

The following welding processes are commonly used:
- submerged arc automatic welding;
- carbon dioxide or argon semiautomatic welding (MAG and MIG, respectively);
- manual shielded-arc process;
- argon manual welding (TIG).

The most extensively used processes for structural work are the automatic welding, for skin plates and hoist drums, and the MAG welding. Other parts of the structure are welded with manual shielded-arc process.

Fig. 14.1 Segment gate leaf shop assembly with jigs (ALSTOM)

14.1.8 FINISHING

Once the welding is over, the fabricated members are cleaned and straightened. Temporary attachment devices (clamps, wedges and lugs), dirt and weld splashes are removed. Disc grinders and sanding machines are used for removal of attachment devices and weld spatters. Structure straightening is made with torches.

14.1.9 PRE-ASSEMBLY

This step consists in the simulation of the final conditions of the parts or blocks, to give an overall view of the assembly. It usually takes place after machining. Sometimes, however, it should be made prior to machining, e.g., the pre-assembly of segment gates necessary for marking the gate hinge axis in the trunnion boss. The sealing frame of sluice gates is completely pre-assembled (Fig. 14.2). Sliding and wheel tracks, guides and counter-guides need not be completely assembled, it being enough to pre-assemble the various elements, in pairs, for checking of field joints preparation. Fixed-wheel gates are usually pre-assembled in the horizontal position, with all accessories (wheels, shafts, seals and so on). Whenever possible, these gates and stoplog panels are vertically lifted by the suspension points, so as to check the center of gravity position. Segment gates are pre-assembled complete, in the vertical position, whenever possible.

An overall verification of the gate dimensions is conducted in the pre-assembly phase. Positioning and alignment of the wheels of fixed-wheel gates are checked. Field assembly marks are also made in this phase.

Fig. 14.2 Pre-assembly of sealing frame and wheel tracks of fixed-wheel gate (Müller)

14.1.10 MACHINING

The most used machines in the machining of gates are shapers, boring machines, lathes, milling and drilling machines. Shapers or boring machines are used for machining of wheel and sliding tracks. Shapers usually machine seal seats of stoplogs and fixed-wheel gates, but the same result can be attained with boring machines. The holes of the main bushing housing of segment gates and, also, the holes of the wheel shafts of fixed-wheel gates are machined by the boring machine. In mechanical hoists, the boring machine is used for machining of the base shims, while drums, shafts and pulleys are machined in lathes. Minor services of machining are performed in milling machines, planers and drilling machines.

Fig. 14.3 Machining of wheel track (KRUPP)

Fig. 14.4 Machining of wheel pin holes of the Mirorós dam fixed-wheel gate

The machining of the seal seats and the lower edge of the skin plate presents a number of advantages that must be taken into account, such as greater manufacturing accuracy, reduction of the assembling time due to the greater facility in obtaining the required tolerances and the reduction of leaks through the seals.

14.1.11 MECHANICAL FIT-UP

Mechanical fit-up is carried out in fixed-wheel, segment and slide gates, hoists, lifting beams and by-pass valves. Mechanical fit-up of wheels, shafts and gate leaf is performed on fixed-wheel gates. In segment gates, the following mechanical adjustments are made: trunnion with the gate pin and gate boss with the bushing and with the pin.

In the assembly of mechanical hoists, the fit-up starts with the drum and the drum gear, then with the bearings, shafts and the base frame. Thereafter, the mechanical fit-up of the pinions, speed reducers, motors and brakes is done. Previous to this phase, a sort of sub-assembly is accomplished, as in the case of the drum with the final gear; speed reducer with the coupling in the input shaft and with the final reduction pinion in the output shaft; motors with the couplings and sometimes, brake pulleys.

14.1.12 ANTICORROSIVE PROTECTION

After manufacture, all unfinished ferrous metal surfaces receive an anticorrosive protection corresponding to the working conditions to which they will be subjected (contact with water, weather exposure, aggressiveness of the environment, mechanical friction and so on). Usually, the customer specifications indicate the type of surface preparation and the paint system to be applied.

The metal surfaces finished and subject to corrosion are protected after manufacture by a heavy coat of varnish or other equivalent material readily removed by commercial solvents. Stainless steel parts, non-ferrous metal or galvanized parts do not receive any protective treatment. The same applies to metal surfaces to be embedded in concrete.

The painting is carried out either in the workshop after manufacture or in the field. In the latter case, the metal surfaces are usually protected at the workshop with a coat of shop primer, to be removed by sandblast before painting.

All surfaces to be painted should be free from dust, grease, mill scale, oxides and foreign matter. Surface preparation may be either manual or mechanical, through wire brushes, sanding machines or blasting.

The most accepted standard in steel surface preparation is the Swedish standard SIS-05-5900-1967 "Pictorial Surface Preparation Standards for Painting Steel Surfaces", with the preparation degrees shown in Table 14.1.

Table 14.1 Surface Preparation Specifications

Subject	Degrees of surface preparation	
	Swedish Standard No. SIS-05-5900-1967	SSPC Standard (equivalency)
a) Tool cleaning		
- hand tool cleaning	St 2	SP 2
- power tool cleaning	St 3	SP 3
b) Blast cleaning		
- brush-off blast cleaning	Sa 1	SP 7
- commercial blast cleaning	Sa 2	SP 6
- near-white blast cleaning	Sa 2 1/2	SP 10
- white metal blast cleaning	Sa 3	SP 5

A brief abstract of the main features of the various degrees of surface preparation, according to the patterns of standard SIS-05-5900-1967 is presented:

. Hand Tool Cleaning - Specification St 2
Removal of loose rust and loose paint by hand brushing, hand sanding, hand scraping, hand chipping or other hand impact tools, or by a combination of these methods. After cleaning, the surface should have a soft metal glaze.
. Power Tool Cleaning - Specification St 3
Removal of loose mill scale, loose rust, and loose paint with power wire brushes, power impact tools, power grinders, power sanders, or by a combination of these methods. After cleaning, the surface should exhibit a pronounced metal glaze.
. Brush-Off Blast Cleaning - Specification Sa 1
A rapid blast cleaning is made on the steel surface to remove loose rust, loose paint and foreign particles, by the impact of abrasives propelled through nozzles or by centrifugal wheels.
. Commercial Blast Cleaning - Specification Sa 2
Careful blasting is made to remove loose mill scale, loose rust, rust-scale, paint, or foreign matter. If the surface is pitted, slight residues of rust may be found in the bottoms of pits; at least two-thirds of each square inch of surface area should be free of visible residues. After treatment, the surface must present a grayish coloration.
. Near-White Blast Cleaning - Specification 2 1/2
Blasting is kept on for enough time to assure removal of nearly all mill scale, rust, paint, and foreign particles, so that only light shadows, slight streaks or discoloration may be present on the surface. Residues are removed by vacuum cleaner, clean and dry compressed air or brushing. At least 95 percent of each square inch of surface area should be free of all visible residues. The surface must present a light gray color.
. White Metal Blast Cleaning - Specification Sa 3
Careful blasting for complete removal of all mill scale, rust, paint, or foreign matter. The final removal of residues is effected through clean and dry compressed air or brushing. After cleaning, the surface should present a gray-white, uniform metallic color, without shadows or streaks.

In general, the durability of an anticorrosive protection system is directly proportional to the final thickness of its dry film. The number of coatings to be applied depends on the roughness of the surface to be protected, the thickness and hardness of the paint and the degree of aggressiveness of the environment to which the equipment is subjected.

Technically, a protection system is considered reasonable to good when the quantity of coats is such that the dry film final thickness is equal to or more than three times the maximum height of profile of the anchor pattern produced on the surface. In the particular case of metal surfaces subjected to white metal blast cleaning, a maximum height of profile of about 40 to 50 micrometers is produced. Therefore, the protecting film final thickness should be from 120 to 150 micrometers. For surfaces subjected to severe mechanical friction, a minimum thickness of 300 micrometers is recommended. Paints with hard film (epoxy, for example) give good protection with small thickness.

Among the paint systems commonly employed in the anticorrosive protection of hydromechanical equipment, one should mention especially:

. Zinc-rich epoxy paints result in hard and thick films (from 60 to 70 micrometers), with great abrasion resistance and good cathodic protection. They are commonly used as bottom coating.
. High build epoxy paints give watertight films with great mechanical resistance and adherence to the substratum. High build epoxy coatings provide long-term protection in situations involving fresh and salt water and may be applied in coats with a minimum dry film thickness of about 200 micrometers. The dry film presents a very hard surface, with a slightly gloss appearance. High build epoxy paints should not be applied in potable water reservoirs and systems.
. Acrylic polyurethane based paints give hard films with satisfactory resistance to water and weather. Due to the small film thickness (from 50 to 75 micrometers), it may require two coats to get adequate protection. Color finishes are available.
. Alkyd resin paints result in relatively hard films with good abrasion resistance and excellent weather resistance. Generally, this paint system is not considered satisfactory for continuous water immersion and should not be applied over bituminous or coal tar based paints. Colored finishes are available.

The above paints are applied either by brushes, roller or high-pressure spray. Recommendations of the paint manufacturer should be strictly followed.

According to the equipment exposure to the action of water and the weather, the following anticorrosive protection systems may be suggested:

a) Permanently submerged equipment (embedded parts, bottom outlet gates, intake gates, for example) or temporarily submerged and exposed to weather (stoplogs and lifting beams, for example):
   - near-white blast cleaning (Specification Sa 2 1/2);
   - one coat of zinc-rich epoxy paint, minimum dry film thickness of 60/75 micrometers;
   - two finish coats of high build epoxy paint, each coat with minimum dry film thickness of 150 micrometers.

b) Equipment with one side in permanent contact with water and the other exposed to weather (spillway gates, for example): the submerged side should be protected as described in item *a* above. The non-submerged side should be protected as follows:

- power tool cleaning with brushes or sanding discs (Specification St 3), or commercial blast cleaning (Specification Sa 2);
- two coats of epoxi based paint with micaceous iron oxide pigment, each coat with a minimum dry film thickness of 75 micrometers;
- one finish coat of acrylic polyurethane based paint, with a minimum dry film thickness of 60 micrometers.

c) Non-submerged equipment exposed to weather (hoists, position indicators, fluid power units and so on):
- power tool cleaning with brushes or sanding discs (Specification St 3);
- one coat of zinc phosphate pigmented paint, alkyd resin vehicles, with a minimum dry film thickness of 40 micrometers;
- two finish coats of alkyd resin paint, each coat with a minimum dry film thickness of 50 micrometers.

14.1.13 INSPECTION

The inspection takes place during the whole manufacturing process. Steps, phases and points subjected to inspection are shown in the production flowcharts. An inspection program is developed before starting the manufacture; it lists and describes in detail the main activities such as raw material control, welders and welding process qualification, dimensional and visual control, and destructive, nondestructive and functional tests of the equipment.

The acquired materials are inspected on reception. Checking of the conformity of the characteristics of the supplied material with the values required by the manufacture drawings, material lists or buying orders is made through the analysis of the quality certificates. Forged and cast parts are inspected at the supplier's facilities. Plates over 19mm thick are inspected by ultrasonic examination.

Welders and welding processes are conducted in the presence of the inspector and must follow the recommendations of the AWS (American Welding Society).

Visual and dimensional controls are made in the various steps of manufacture. In the visual inspection, the position and assembly sequence of parts, the edge preparation, the size, length and location of welds, and the overall finishing are checked, among others.

Visual inspection of welds allows detection of leg or throat undersize, excessive convexity, undercutting, overlapping. Check should be made to detect that no weld has been omitted or that no unspecified weld has been made. Dimensional control is base on approved customer's drawings and the required tolerances.

Nondestructive tests are very useful in the quality control of manufacture. Their main features are:

a) Magnetic particle examination detects any superficial crack in the welds and reveals defects up to 3 mm deep. In multi-layer welds, each layer must be examined to inspect the weld properly.
b) Liquid penetrant examination reveals only surface cracks. It is quick and easy to apply.
c) Ultrasonic inspection is quick and precise, but requires personnel competent to use the equipment. It reveals inner cracks.

d) Radiography is used in butt welds (skin plate joints, e.g.). It shows weld defects that are not detected by visual inspection such as lack of fusion or penetration, porosity, or slag inclusion, and allows permanent record of the defect.

Destructive tests (tension, bending, impact, macrography etc) are used in the determination of the physical properties of cast or forged parts, in the re-qualification of welding procedures or, further, in case of doubt as to the quality of welds or raw material.

Functional tests of equipment are conducted in the factory, in the presence of the inspector. The main tests are:

a) Wheels and rollers: manual rotation;

b) By-pass valves:
- opening of the plug disc by action of the lifting beam weight on the driving lever;
- closing of the plug disc after removal of the lifting beam.

c) Lifting beams:
- movement of hooks (opening and closing);
- operation of levers and counterweight. The predominance of the counterweight action in the opening and closing of hooks must be checked;
- engagement and disengagement of hooks in the gate panel lifting lugs.

d) Balancing of lifting beams and gates.

Fig. 14.5 Shop test of the Mornos dam gate (MAN)

14.1.14 MANUFACTURING TOLERANCES

The equipment main dimensions are measured and recorded during the shop pre-assembly. Dimensions may vary within the tolerance range set out in the approved manufacture drawings, the customer specifications, or the manufacture standards.

Tolerances may be of shape (flatness, straightness, cilindricity etc), orientation (parallelism, perpendicularity, angularity, verticality etc), or of position (symmetry, concentricity, coaxiality etc). Tolerances of mechanical parts, such as shafts, pins, bushings, wheels, are usually established as a function of the required accuracy degree, the class of fit (tight, loose, free etc), or the technique of assembly (thermal, press, manual etc).

Table 14.2 Permissible Variation in Linear Dimensions For Welded Structures (DIN 8570 - Part 1)

Degree of accuracy	Nominal dimension range (mm)									
	Over 30 up to 120	Over 120 up to 315	Over 315 up to 1000	Over 1000 up to 2000	Over 2000 up to 4000	Over 4000 up to 8000	Over 8000 up to 12000	Over 12000 up to 16000	Over 16000 up to 20000	Over 20000
A	±1	±1	±2	±3	±4	±5	±6	±7	±8	±9
B	±2	±2	±3	±4	±6	±8	±10	±12	±14	±16
C	±3	±4	±6	±8	±11	±14	±18	±21	±24	±27
D	±4	±7	±9	±12	±16	±21	±27	±32	±36	±40

A permissible variation of ±1 applies to dimensions up to 30mm.

For linear dimensions of welded structures, whose tolerances have not been indicated in the manufacture drawings, the values recommended by the DIN standard 8570 - Part 1, reproduced in Table 14.2, may be used, according to the desired accuracy degree. Results obtained with accuracy degree B are usually satisfactory. Different degrees of accuracy may be considered in the same structure.

The following are some tolerances that can be applied to steel gates with rubber seals and dimensions up to 10 m by 10 m. For gates exceeding those values, tolerances referring to lengths can be readjusted as follows:

$$\varepsilon = \varepsilon_0 (1 + L/10)/2 \qquad\qquad (14.1)$$

where:

ε = dimensional tolerance of the member;

ε_0 = dimensional tolerance for a length of 10 m;

L = gate span or height, in meters.

Example: In a fixed-wheel gate 15 m wide by 8 m high, the allowable tolerances for the dimensions measured in the span direction are increased as follows:

$$\varepsilon = \varepsilon_0 (1 + L/10)/2 = 1.25 \, \varepsilon_0$$

In this case, the tolerances relative to the dimensions measured in the height direction do not change, for $h = 8$ m < 10 m.

For very flexible elements (slender vertical guides, for example) that do not attain the required tolerances, it is sufficient to check in the fabricating shop that these tolerances can be achieved in the field assembly through adjustment of the anchor bolts or use of wedges and rods.

Embedded Parts

a) Sill
Sealing surface flatness, measured on the contact range:
± 1 mm/m and ± 1.5 mm in the total.
b) Seal seats
Sealing surface flatness, measured on the contact range:
± 1 mm/m and ± 1.5 mm in the total.
c) Wheel track (or slide track)
Flatness measured on the centerline of the contact range, in the region where the gate is subjected to water pressure:
± 0.5 mm/m and ± 1.5 mm in the total.
d) Distance between the plans of the seal seats and of the wheel (or slide) track:
± 0.5 mm.
e) Side guides and counterguides
Flatness measured in the centerline of the contact range:
± 1 mm/m and ± 2 mm in the total.

Gate Leaf

a) Flatness of the lower edge of skin plate:
± 1 mm/m and
± 1.5 mm in the total.
b) Flatness of end vertical girders of stoplogs and slide gates:
± 1 mm/m and
± 1.5 mm in the total.
c) Flatness of the seal base plates:
± 1mm/m and
± 1.5 mm in the total.
d) Coplanarness of the seal base plates (flat gates):
± 1.5 mm.
e) Distance between the sliding (or tread) surfaces of the side shoes (or guide wheels), measured in the direction of the span:
± 2 mm.
f) Distance between the centerlines of the main wheels (fixed-wheel gates), measured in the direction of the span:
± 2.5 mm.
g) Distance between the centerlines of the end vertical girders (stoplogs):
± 4 mm.
h) Distance between lifting lugs (stoplogs):
± 3 mm.

i) Variation of the segment gate radius, measured in the skin plate reference face, in the region where the seals are attached:

   ± 6 mm.

Lifting Beams

a) Distance between the tread (or sliding) surfaces of the guide wheels (or side shoes), measured in the direction of the span:

   ± 2 mm.

b) Distance between hook centers:

   ± 2 mm.

c) Vertical distance between the beam stop and the hook contact face with the suspension pin or plate:

   ± 1.5 mm.

14.2 TRANSPORTATION

Due to its importance, transportation of gates must be carefully planed and executed. The work starts with the choice of the access roads and the transportation means to the site, it being desirable to avoid transhipment. Very wide equipment requires the presence of forerunners in roadways. Very tall parts may require the temporary relocation of bridges crossing over the road or the use of side detours or side roads close to overpasses. In certain cases, it is necessary to combine various means of transportation to arrive at the site, such as trucks, trains, barges and ships.

After manufacture, the equipment is prepared for the transportation to the field. Screws, nuts, anchor bolts, pins, tubes, fittings, seals, are identified by tags containing the drawing and part numbers. These components, as well the electrodes for field welding and the finishing paints are then arranged in bags or wooden crates externally identified for transportation. Machined parts are protected against corrosion and mechanical damages by means of grease, varnish, wood pieces or other adequate material. A watertight plastic cover or a tarpaulin protects mechanical or electrical equipment and components susceptible to be damaged by rainwater. The hydraulic system pipes conveyed separately are internally protected with corrosion inhibitors and plugged at the ends.

Large equipment such as wheel tracks, guides, gate leaves and lifting beams do not need close packing and are conveyed in trailers or open trucks, duly choked, stacked and tied, to avoid deformation. Equipment should preferably be totally contained in the base of the transporting vehicle. Equipment or parts exceeding the limits of the base of the transportation vehicle may be transported with sections in cantilever, provided adequate measures are taken as to the balance of the set and undue deflections. In special cases, it is usual to transport long equipment or parts supported on a trailer, provided this has been previously agreed between customer, transporter and manufacturer. In these cases, the equipment or part serves as the link to the traction element of the trailer.

During transportation, the accumulation of rainwater in the equipment or parts placed over the transporting vehicle should be avoided, so as not to impair the balance of the set vehicle/load, or cause overload on the vehicle.

14.3 FIELD ERECTION

14.3.1 ERECTION INSTRUCTIONS

The field erection of the gate components usually follows the sequence: embedded parts, gate and hoist. The erection is carried out according to the instructions, standards and procedures prescribed in the specifications or supplied by the manufacturer. These documents are usually gathered in a single volume, called Erection Manual. The manual should cover the following information:
. list of drawings and other reference documents;
. erection sequence;
. instructions for handling and indication of the suspension points of large parts;
. dimensions and weight of the main parts;
. instructions for placing, alignment, leveling and fixation of parts;
. marks for assembly;
. instructions for concreting of embedded parts;
. field junctions and welds to execute (including the recommended type of electrodes).

The manual may also contain the time schedule, the erection staff and a list of the auxiliary equipment required for the erection (cranes, hoists, jacks, turnbuckles etc).
    The erection can be integrally done by the gate manufacturer or, more often, by others, under his supervision. For the erection one should not include any field machining, welding preheating or stress relieving.

14.3.2 ERECTION OF EMBEDDED PARTS

The embedded parts are assembled in block-outs built in the concrete, connected to the first stage concrete anchor plates through anchor bolts and then concreted. Installing the embedded parts directly in the first stage concrete requires a rigid attachment to prevent their movement during the pouring and cure of the concrete.
    Erection of embedded parts is usually preceded by a topographic survey and cleaning of the working area. To facilitate erection, the main reference points and lines, such as axes, elevations and centerlines should be previously marked on the concrete with paint, metal wires or pins. Prior to start the erection, slot dimensions, location of first stage anchor plates, and the erection reference points and lines are checked according to the construction drawings, in the presence of the owner and the gate supplier supervisor.
    The usual sequence of the embedded parts erection is:

a) sill beam;
b) lower elements of the vertical parts (wheel or sliding tracks, counter-guides and side guides);
c) lintel beam (for sluice gates only);
d) top elements of the vertical parts.

As an exception to the rule, one may mention the embedded parts erection of the spillway segment gates of the Itumbiara Power Plant, where, to observe a particularity of the construction time schedule, the erection and concreting of the lateral embedded parts of some spans started before the conclusion of the crest concreting. In these spans, even the gates were assembled before concreting the crest (Figures 14.6 and 14.7). Afterwards, the crest was concreted and the sill beams installed.

Fig. 14.6 Field assembly of the spillway segment gate at the Itumbiara Power Plant before completion of the crest

Fig. 14.7 View of the temporary support for field erection of the Itumbiara spillway segment gate

In order to reduce the erection time and facilitate fieldwork, it is common to transport and assemble the sealing frame of gates of small dimensions in a single block. In the sill beam assembly, the following steps are followed:
. pre-assembly of the anchor bolts in the beam;
. beam positioning within the block-out and welding of the anchor bolts in the first stage anchor plates
. sill beam alignment and leveling through the anchor bolt nuts.

After dimensional checking and liberation of the erection by the inspector, block-out cleaning and concreting are carried out. The two ends of the sill beam are kept free from concrete, so as to allow the laying of the vertical elements and their junction with the sill beam.
In assembling the vertical elements, the procedure is as follows:
. pre-assembly of the anchor bolts in the element to erect;
. positioning of the vertical element within the lateral block-out. The embedded parts are usually supported on metal shims, made from scrap, and leveled by means of jacks;
. tack-welding of the end anchor bolts to the anchor plates;
. bolt connection of the sill beam with the vertical elements;
. welding of the remaining anchor bolts to the anchor plates;
. leveling, vertical alignment and correct positioning of the element with respect to the flow direction, through the anchor bolt adjusting nuts.

The concrete forms are placed after the dimensional checking and block-out cleaning. The concrete is then poured in layers with a maximum height of 1.5 m to 2.0 m. Care should be taken to not displace or damage the parts being assembled. Also, concrete pouring should take place with moderate speed, so as to avoid great pressure of the concrete on the embedded parts. The concreting is interrupted at about 0.5 m from the upper end of the vertical element, so as to permit its connection with the adjacent upper element.

14.3.3 ERECTION TOLERANCES OF EMBEDDED PARTS

The positional adjustment of embedded parts is done with the anchor bolt nuts and their checking is accurately made with the help of measuring apparatus and devices such as optical level, plumb line, metal tapes etc.
The dimensional and positional checking of the embedded parts is made in the presence of the erection inspector and supervisor. Release for concreting of the embedded parts takes place only after verification that the erection tolerances indicated in the approved drawings has been accomplished.
The shape tolerances of embedded parts indicated in item 14.1.14 may also be applied to the final field erection, in the absence of specific values. Further erection tolerances are:

a) Maximum leveling run-out of sill beam: 4 mm.
b) Wheel (or sliding) track span, measured in the centerline of the contact band: ± 3 mm.
c) Side seal span, measured in the centerline of the contact band: ±4 mm.
d) Span of side guides and counter-guides, measured in the centerline of the contact band: ± 3 mm.

e) Face to face distance between wheel (or sliding) track and the counter-guide: ± 2 mm.

f) Distance between the wheel (or sliding) track and the centerline of the side guide: ± 2 mm.

g) Verticality of the wheel (or sliding) tracks, side seal seats, side guides and counter-guides:
   - in the vertical plane parallel to the flow: ± 1.5 mm;
   - in the plane normal to the flow: ± 2.0 mm.

h) Coplanarness of the seal frame: ±1.5 mm.

14.3.4 GATE ASSEMBLY

Small gates, lifting beams and stoplog panels are completely assembled in the workshop, transported and installed in their final place without requiring any additional erection activity in the field.

Gates with height exceeding the transport limitations are horizontally subdivided in elements with maximum height around 3 m. The various elements of fixed-wheel gates have their wheels and shafts assembled in the workshop and are conveyed in trailers or trucks to the site. Their placing in the slots is carried out in sequence, starting with the lower element, through wheel cranes or gantry cranes. The lower element is placed in the slot and supported on horizontal dogging beams located in the maintenance chamber deck. The intermediate element is then placed on the locked element and their junction made, usually with bolts. In this phase the seal is also installed between the elements. The above procedure is repeated with the remaining elements until complete assembly of the gate.

The gate is then lifted slightly by the gantry crane for removal of the support beams. The gate is lowered until its top coincides with the maintenance chamber deck, where locking with the dogging beams takes place. The next steps are the assembly of the gate hoist and its connection to the gate.

The field erection of a weir segment gate usually follows this sequence:
. trunnions and bearings;
. radial arms;
. gate leaf;
. seals.

The erection of the main bearings is a complex operation and requires great accuracy. The positioning of the pivoting axle in the vertical plane, the leveling and coaxiality of the bearings must be carefully controlled. The radial arms are fastened to the bearing housings by bolts or welds, depending on the design. In large segment gates, the radial arms located in either side of the gate do not form a single block and are assembled by parts, starting with the lower arm.

The erection of the gate leaf, when formed by various elements, starts with the lower element, which is placed on the sill through cranes and connected to the lower radial arms. The gate leaf elements are kept in the assembly position by means of metal staying, turnbuckles and, in some cases, with angle pieces tack-welded to the skin plate and the side guides (see Fig. 14.8). The skin plate elements are joined together by welding. Scaffolds are usually employed in this operation.

Fig. 14.8 Temporary device used for the field erection of spillway segment gate leaf

The radial upper arms of large segment gates are installed only after assembly of the corresponding element of the gate leaf, to which they will be joined. Installation of seals is made after completion of the gate assembly and painting.

14.4 ACCEPTANCE TESTS

After assembly, gates and hoists are subjected to a series of acceptance tests programmed by the owner for assessment of their performance. The acceptance tests are classified in preliminary and operational.
Preliminary tests are usually conducted under dry conditions, soon after completion of the assembly, and serve to simulate the gate operation. These tests consist of:

a) Stoplogs
- free sliding of panels along the embedded parts;
- interchangeability of panels and gates;
- lifting beam operation (installation and withdraw of panels);
- operation of panel dogging devices.

b) Spillway service gates
- free opening and closure of gates;
- maximum lifting course of the hoist;
- gate opening and gate stop in the intermediate positions;
- gate stop in any position of the course;
- total closing of the gate, in continuous operation, from the totally open position;
- functioning of limit switches, controls and signaling;
- operation of the position indicator;
- operation of the hand drive mechanism.

c) Emergency gates of hydraulic turbines
- free opening and closure of gates;
- maximum lifting course of the hoist;
- gate opening and gate stop in the cracking position;
- complete opening of the gate, in continuous operation, from the cracking position;
- total closing of the gate, in continuous operation, from the totally open position;
- functioning of the limit switches, controls and signaling;
- operation of the position indicator;
- operation of the hand drive mechanism.

Operational tests are conducted after filling the reservoir and serve to check the gate performance under actual conditions. Generally, the following tests are conducted in this phase:

a) Stoplogs
- leakage through gate seals;
- proper functioning of the by-pass valve.

b) Spillway service gates
- leakage through gate seals;
- raising and lowering speeds;
- gravity closure of the gate.

c) Emergency gates of hydraulic turbines
- leakage through gate seals;
- gravity closure of the gate, with the generating unit operating at maximum power.

In gates operated by hydraulic hoists, the following items are also checked:
. correct functioning of the end-of-stroke damping;
. maximum pressure in the hydraulic system;
. interconnection of the adjacent pump units (when it applies).

Emergency gates should be judiciously tested, for they are the ultimate protection of the generating unit in the event of failure in the regulation system, rupture of the guide vanes shearing pins, or rupture of the penstock. In the reference [1] a detailed description of the recommended tests for commissioning of emergency gates is found.

After completion of the field tests with satisfactory results, the Preliminary Acceptance Document is issued, starting then the guarantee period, during which the supplier is bound, without charge for the owner, to repair all defects caused by design, materials and manufacture. At the end of the guarantee period, if the performance of the equipment is considered satisfactory, the Final Acceptance Document is issued.

REFERENCE

1. Matos, G., Cyranka, P. and Cavalcante, A.F.: Commissioning of Emergency Gates of Hydraulic Turbines (in Portuguese), *VII SNPTEE*, Brasília, 1984.

Chapter 15

Trends and Innovation in Gate Design

15.1 LONG-SPAN GATES

Projects for flood and tide protection require the design of unusually large gates. In developing projects with very wide gates, the designers are obliged to concentrate their effort on obtaining reliable gates, low operating forces and correct synchronization between the lifting mechanisms.

Long-span gates are very common in Japan and some countries adjoining the North Sea (Germany, Netherlands and the United Kingdom), either for tidal protection or for damming rivers for agricultural purposes. The solutions encountered by designers for the development of long-span gates are varied and highly creative.

To protect the country's population from seawater flooding the Dutch Government passed the Delta Act in 1955, which launched the Delta project, following the disastrous flooding of 1953. This project proposes the closure of the main tidal estuaries and inlets in the southern part of The Netherlands, without limiting navigation access to cities of Rotterdam and Antwerp, and without a negative impact on the environment.

One of its most challenging projects was the Eastern Scheldt storm surge barrier, located in the 50 m-deep, 5 km-wide river mouth, and involving 62 large-span slide gates, which are to be closed at extreme high tides. In normal weather conditions the gates remain open. The gates have a span of 41.4 m and their height varies from 5.9 m to 11.9 m as the flow profile of the barrier roughly follows the bottom profile of the tidal channels. Individual gate weight varies between 2900 and 5100 kN. Two hydraulic cylinders operate each gate. The water retaining plating is composed of cylindrical segments at the central part and of orthotropic plates at the sides. The main girder system consists of two or three horizontal truss type girders, depending on the height of the gates. The horizontal forces acting on the gates are transferred to the piers by vertical end beams located in the recesses of the piers. The end beams are fitted with flexible guide shoes, which compensates for offset of the piers, bending of the gate, and gives a pre-loading of the gates between the guides. The pre-loading prevents the gate knocking against the guide strips due to horizontal wave action.

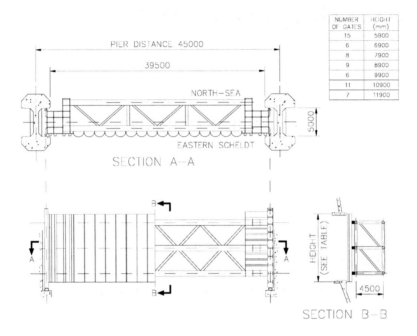

NUMBER OF GATES	HEIGHT (mm)
15	5900
6	6900
8	7900
9	8900
6	9900
11	10900
7	11900

Fig. 15.1 Eastern Scheldt storm surge barriergate

Another major element of the Delta Project is the Hartel canal barrier, located in the area of the city of Rotterdam and the Europoort. This canal connects the open sea with the inland canals and rivers. Two gates with spans of 98 m and 49 m, and weighing 6180 kN and 2256 kN, respectively, will close the Hartel canal in the event of a flood. Each gate is operated by two 21.2 m-stroke hydraulic cylinders, which operate in a synchronized mood.

Fig. 15.2 Hartel canal gate, powered by ABS hydraulic cylinders of Rexroth Hydraudyne B.V.

Large radial gates are often used in Germany and Netherlands to form a barrier against high water from the sea. These gates, with spans equal to or larger than 40 m, provide protection against saltwater intrusion into waterways and are used to create freshwater basins.

A barrier was constructed in the mouth of the New Waterway, near the city of Rotterdam, in The Netherlands. During storm surges, the barrier will close off the 360 m-wide waterway; the rest of the time it will remain open, so as not to obstruct navigation. The barrier consists of two identical radial gates, one on each side of the waterway. Each structure is 22 m-high and has a 246 m-radius curved retaining wall, connected by two 246 m-long steel truss arms to a pivot on the riverbank.

Fig. 15.3 New Waterway double radial gates

The semi-submersible gates are free-floating structures, with varying degrees of lateral, vertical and horizontal movement provided by massive ball joints. They are constructed in curving dry docks on each bank, which will open and flood when the river level rises. The dock is closed by steel box gates, which allow the dock to be pumped dry for maintenance. To close the barrier, the gates are floated into the river and then ballasted until they sink onto concrete sills. To lower the gates, water is let into built-in chambers by opening the valves. This is done in two stages, the gate being lowered to just above its final resting position to allow fast water flow to scour silt and sand off the top of the sill, and then, when clear contact can be made, dropping the gate down fully. To raise the gates, water is simply pumped out the chambers and the barrier floats to the surface ready to be maneuvered back into the dry dock. The ball-joints, at the pivots, are designed to withstand hydraulic and load forces around 370 MN for each gate. A single ball joint weighs 6670 kN. It has a steel plate core to which spheroid cast steel elements are attached. The ball revolves in eight concave elements, also constructed from cast steel, which are attached to the concrete foundation. Ten meters in diameter, the joint is accurate to one millimeter.

On the riverbank, a triangular concrete base weighing 510 MN surrounds the ball joint.

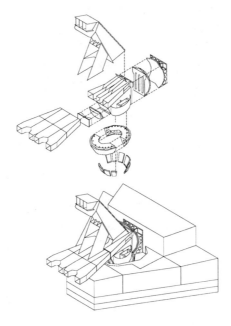

Fig. 15.4 New Waterway radial gate ball joint

During the heaviest storm conceivable, the joint may move 200 mm backward during closure of the barrier, but will subsequently move back halfway. The forces acting on the joint swing through almost 90° as the barrier moves out of the dock, from vertically downwards when the gate is open to almost horizontal when closed [1, 2, 3].

Radial gates of unconventional design have been used at the Thames Barrier to protect London from the North Sea tidal flooding. Designers considered both vertical lift and drum gates before deciding on their own *rising* radial gate design for the six navigational openings, four 61 m-wide and two 30.5 m-wide. The gates are able to rise 18 m, yet rotating down into recesses in the riverbed that are only a fraction of their vertical height, providing a clearance of at least 9 m below mean water level. They were designed as box girders and mounted on steel disks with counterweights that rotate on trunnions located in the piers. The trunnion shafts are mounted on 2200 mm-diameter self-lubricated spherical bearings. Each gate weighs about 31 MN and supports a hydraulic load of 76.7 MN (horizontal component only). The same concept of "rising" radial gate was used in the design of the 24 m-wide Freudenau shiplock segment gate, on the Danube river.

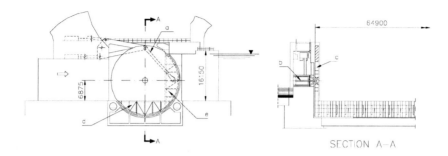

SECTION A—A

Fig. 15.5 Thames barrier radial gate
(a) lever; (b) bearing; (c) counterweight; (d) gate opened; (e) gate closed

Fig. 15.6 Downstream view of the Thames barrier radial gates

Top-hinged flap gates are widely used in Germany to protect navigation channels against storm floods. These gates, with spans varying between 30 and 42 m, are normally kept in the open position.

Visor gates were developed to control tidal effects and permit the passage of vessels when raised. Visor gates with 48 m and 56 m span have been installed in Netherlands and in Japan.

Table 15.1 shows the main dimensions of some gates with spans of more than 30 m.

Table 15.1 Outstanding Examples of Large-Span Gates

Year	Project	Span (m)	Height (m)	Head (m)	Gate type
1995	New Waterway	360.0	22.0		Double-radial with vertical hinges
	Volga's delta	110.0	12.9	4.2	Fixed-wheel
	St. Pantaleon	100.0	3.7	3.7	Flap
	Hamilton	91.0	8.5		Drum
1982	Thames	61.0	20.1	8.4	Raising radial
1970	Kizugawa	57.0	11.9	10.9	Visor
1967	Haringvliet	56.5	10.5		Segment
1949	Hemelingen	54.0	4.6	4.6	Sector
1974	Kitagami	50.0	6.1	6.1	Fixed-wheel with flap
1960	Hagenstein	48.0	7.0		Visor
1949	Donzère-Mondragon	45.0	9.1		Segment with flap
	Idaho Falls	45.0	3.0		Flap
	Hahlen	43.9	4.3	4.3	Fixed-wheel
1972	Arakawa	43.6	5.0	5.0	Fixed-wheel
1976	Stör	43.0	13.0		Segment
	Artlenburg	42.0	9.6	9.6	Top-hinged flap
1970	Artlenburg	42.0	5.0	4.5	Fixed-wheel
1985	Scheldt	41.3	11.9	6.2	Slide
1967	Vilyui	40.0	14.0	13.2	Segment
	Eider	40.0	11.1		Segment
	Vamma	34.0	7.0		Segment
	Billwerd Bucht	34.0	12.3		Top-hinged flap

15.2 HIGH-HEAD GATES

The continued development of hydraulic works of ever-increasing size have led designers and gate manufacturers to look for solutions for complex engineering and associated problems. Intense research and experimental work have been carried out in Japan, China and the former USSR, in the development of gates for the passage of discharges of up to 2000 m^3/s at heads of up to 200 m.

High-pressure gates are normally placed in tandem, the upstream gate being used as the emergency or service gate, and the downstream one as the regulating gate. Segment or slide gates are generally selected as regulating gates. Slide gates are less susceptible to vibration in partial openings because of the large friction forces developed between the sliding surfaces, but they need more powerful hoisting mechanisms. The emergency and main regulating slide gates of the low-level outlets of the Inguri arch dam, in the Republic of Georgia, are designed to close a 5 m-diameter conduit under a head of 181 m and are operated by 8.8 MN lifting capacity hydraulic hoists. The lower edge of the gate was designed with a curvilinear outline, which provided the necessary training of the jet beneath the gate to keep it from striking the downstream faces of the slots, under partial openings of the gate (Fig. 15.7). The emergency gate has a ring follower, which allows the side grooves to be tightly closed when the gate is open. At the entrance of the conduit there is a buoyant guard gate, which is lowered along guides on the upstream face of the arch dam [4].

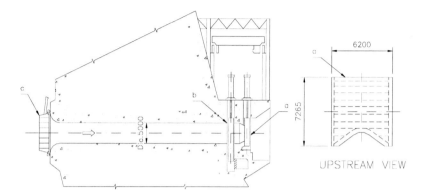

Fig. 15.7 Inguri dam outlet gate, head 181 m
(a) regulating slide gate; (b) ring-follower gate; (c) floating gate

Segment gates are suitable for large discharges in high-head installations since no slots
are required. Compared with vertical-lift gates, the segment gates provide a better
hydraulic configuration and release water smoothly at all openings. However, this type
of gate is not recommended for the outlet works of thin arch dams because of its large
size and the heavy concentration of the load on the trunnions. One of the most
important refinements introduced in high-head segment gate design was the
development of eccentric trunnions (Fig. 15.8). In these gates, seals are placed on the
four edges of the upstream face of the skin plate like a picture frame.

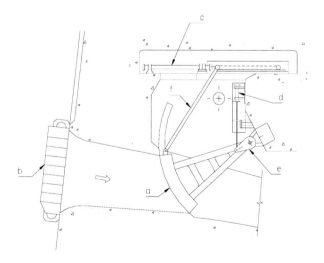

Fig. 15.8 Ohdo dam outlet conduit radial gate with eccentric trunnions, 60 m head.
(a) regulating gate; (b) emergency gate; (c) gate hoist; (d) eccentric trunnion hoist; (e) eccentric
trunnion; (f) gate hoist rod

The eccentric trunnion permits a gap to be created between the seals and seal seats before opening the gate, so that only moments caused by the gate weight and hinge frictional forces have to be overcome. To open the gate, a special hoist first rotates the eccentric shaft and moves the gate backward with little effort. When the gate closes, the steps are reversed and the eccentric device makes the gate move forward and press against the seals all along the orifice contour. This device was developed in the 1950s [5] and used with success in the segment gates shown in Table 15.2.

Table 15.2 Segment Gates with Eccentric Trunnions

Year	Project	Span (m)	Height (m)	Head (m)
1972	Tarbela	4.9	7.3	135.6
1986	Longyangxia	5.0	7.0	125.0
1986	Dongjiang	6.4	7.5	120.0
1980	Nurek	5.0	6.0	110.0
1960	Futase	5.0	3.2	69.0
1973	Kusaki	3.2	3.7	65.6

Table 15.3 shows the main features of some sluice gates designed to operate at heads exceeding 120 m.

Table 15.3 Outstanding Examples of High-Head Gates

Year	Project	Span (m)	Height (m)	Head (m)	Hydrostatic load (MN)	Gate type
1964	Beaver	2.0	3.2	285.9	17.91	Slide
1956	Flumendosa Mulargia	2.0	4.0	210.0	16.32	Slide
	Mauvoisin	1.8	3.0	200.0	10.51	Slide
1970	Mica	2.3	3.5	189.0	14.98	Slide
1980	Inguri	4.35-dia		181.0	34.37	Slide
1956	Schwarzach	2.9	2.9	180.0	15.00	Slide
1980	Holen	1.5	3.0	180.0	7.88	Slide
1969	Keban	1.8	3.7	156.1	10.36	Slide
1970	Emosson	1.1	1.8	155.0	3.07	Slide
1980	Ulla-Førre	2.5	6.0	150.0	21.63	Fixed-wheel
1954	Cancano II	2.0	3.1	145.0	8.89	Slide
1973	Tarbela	4.1	13.7	141.0	75.26	Fixed-wheel
1972	Tarbela	4.9	7.3	135.6	46.97	Segment
1964	Sayano-Shushensk	5.0	8.4	135.0	54.20	Slide
1976	Tous	6.6	6.0	133.0	50.50	Fixed-wheel
1986	Longyangxia	5.0	7.0	125.0	64.00	Segment
1986	Donjiang	6.4	7.5	120.0	78.74	Segment

15.3 REFURBISHMENT AND MODERNIZATION OF GATES AND DAMS

When dams are heightened to increase the volume of stored water, the increase in hydrostatic pressure on the submerged gates makes it necessary to re-check the gates and also their hoists, because of the larger operating forces. Also, the rise of the maximum water level may cause: the heightening of existing crest gates or their substitution by higher gates; the installation of new gates above the existing ones; and the need of new gates on top of the dam.

15.3.1 HEIGHTENING OF EXISTING GATES

Heightening of crest gates can be achieved either: by top or bottom gate leaf heightening; by converting the crest gate into an orifice gate by the construction of a *mask or breast beam*; or by crest heightening. Gate leaf heightening increases the load on bearings and anchorages, the weight of the gate and the magnitude of the operating forces. Larger loads on the bearings may result in replacing the bushings as well as reinforcing the trunnion anchorages. Also, the increase of the operating forces may affect the gate hoist capacity.

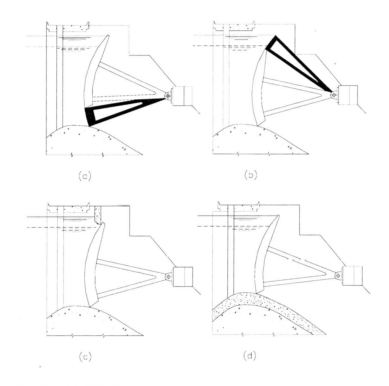

Fig. 15.9 Alternatives for gate heightening
(a) bottom heightening; (b) top heightening; (c) mask beam; (d) crest heightening

- Top heightening. This is accomplished by the installation of a new gate panel on the top of the gate leaf. This new panel can be either a simple extension of the skin plate and vertical beams or a flap gate hinged in bearings mounted on the top of the existing structure. Its main disadvantage is the necessity of general reinforcement of the gate's structure due to the increase of the hydrostatic load.

Top heightening is very simple and was used in the seven 11.5 m-wide by 15.8 m-high segment gates of the Furnas Dam, Brazil (Fig. 15.10), and in the four 24 m-wide by 11 m-high segment gates of the Villalcampo Dam, in Spain (Fig. 15.11), which have been heightened by 1.5 m and 2 m, respectively.

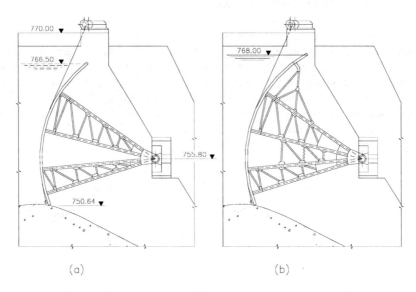

Fig. 15.10 Top heightening of the Furnas dam radial gates
(a) before refurbishing; (b) after refurbishing

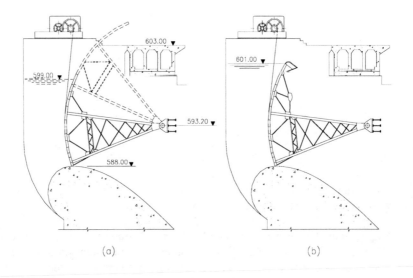

Fig. 15.11 Top heightening of the Villalcampo dam radial gates
(a) before refurbishing; (b) after refurbishing

In May 1988, at the Seven Mile project in Canada, additional sections have been added
to the top of each of the five 15.2 m-wide by 12 m-high spillway fixed-wheel gates,
allowing for an increase of 4.6 m to the maximum operating water level.

The installation of new gates on top of the existing ones was the alternative chosen by designers for refurbishment of the Ilha dos Pombos dam in Brazil. In its spillway, three sector gates 45 m-wide by 7.4 m-high and eight *Stoney* gates were installed in 1924. To permit rising of the reservoir water level, four Stoney gates have already been heightened by 1 m through the top extension of their skin plate, while temporary timber structures were installed on top of the sector gates and the remaining *Stoney* gates. For the sector gates, it was decided to substitute the timber structures by 1.5 m-high reversed segment gates actuated by a pair of hydraulic cylinders.

The segment gates were designed to make contact with the top of the lifted sector gates. The segment gates will be opened for surface spilling or regulation. For large discharges, both gates will be opened.

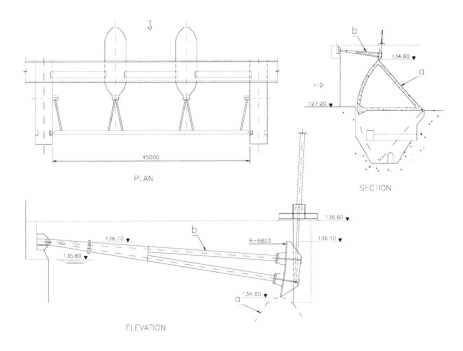

Fig. 15.12 Ilha dos Pombos dam, Brazil
(a) sector gate; (b) top radial gate

- Bottom heightening. This involves the addition of a new gate panel in the bottom of the gate leaf. Loads on the original gate structure elements remain practically unchanged (an exception being made to the bearings in the case of segment gates) and the gate needs no reinforcement. This procedure was adopted in the four 8 m-wide segment gates of the Castro dam, in Spain, which have been heightened from 11 m to 13 m (Fig. 15.13). The plain bronze bushings were replaced by self-lubricated bearings and the anchorages have been reinforced.

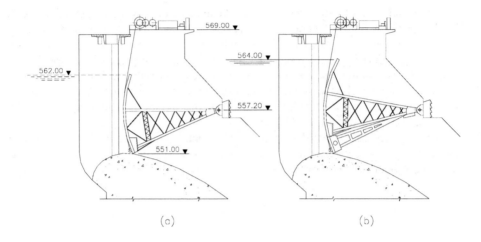

Fig. 15.13 Bottom heightening of the Castro dam radial gates
(a) before refurbishing; (b) after refurbishing

- Mask beam. This consists of the construction of a structural beam just upstream and next to the top of the gate leaf, together with the installation of a top-sealing device between the skin plate and the beam. With the gate closed, the reservoir water level can be raised up to the beam top edge. At the Tucurui dam, in Brazil, provision for the installation of concrete beam masks in the 23 spillway bays has already been made, to transform the 20 m-wide by 21.22 m-high segment crest gates into orifice gates, allowing for a 2 m-rise in the maximum normal reservoir water level. This work was accomplished in 2002.

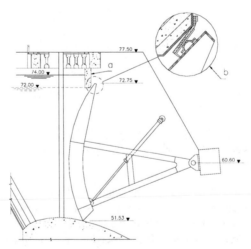

Fig. 15.14 Mask beam of the spillway of the Tucurui Dam, Brazil
(a) mask beam; (b) top seal detail

- Crest heightening. This is achieved by adding concrete to the existing ogee. The hydraulic load remains practically unchanged, and no reinforcement to the gate structure is required. Segment gates need only to turn up on their trunnions. This solution is particularly advantageous when obtaining information on the existing equipment is difficult or unreliable. It is important to note that the amount of civil work is considerable. The reservoir water level has to be lowered below the spillway crest elevation or the water passage should be closed just upstream of the spillway by means of a bulkhead gate.

The reservoir water level of the Cachoeira Dourada dam in Brazil was raised by one meter. The dam has two spillways, each equipped with ten segment gates 16 m wide. The left spillway gates were originally 10.7 m high and the right ones 7.7 m high. It was decided to raise the concrete crest profile while leaving the gates unchanged (Fig. 15.15). Since the drawdown of the reservoir was not recommended, a floating bulkhead located just upstream of each of the spillway bays was used to allow the placement of concrete in dry conditions [6]. The floating gate is linked by means of articulated pins to a horizontal structural beam resting on the top of the lateral piers. Flooding the air-tight chambers causes the gate to turn down to the vertical position. In the closed position, the gate rubber seals contact the concrete face of the crest and the pier noses, enabling the spillway bay to be unwatered.

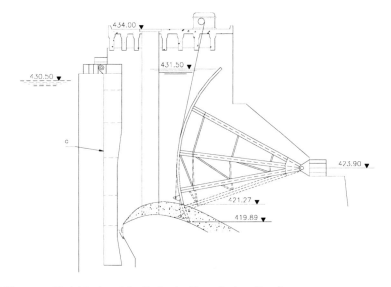

Fig. 15.15 Spillway crest heightening at the Cachoeira Dourada dam, Brazil
(a) floating bulkhead

15.3.2 INSTALLATION OF NEW GATES ON THE TOP OF THE DAM

This case includes dams where the discharge is made through uncontrolled spillways. In such works, the increase of the water level can be accomplished by the installation of either metal gates or inflatable gates.

Inflatable weirs have been successfully installed on the crest of existing dams to increase water head. In hydroelectric installations, they replace flashboards while offering the advantage of simple operation and eliminating the loss of revenue resulting from reduced head after flashboard failure. Inflatable weirs have also been used in pollution control; irrigation; flood control; water supply; the creation of resort lakes; replacing of existing gates; and as tidal barriers to prevent the entry of salt water into estuaries.

The invention of the inflatable weir is attributed to Prof. Mesnager, in France, in 1955 [7]. The new technology had a rapid progress in the USA, notably by the Firestone company, and in particular by Norman Imbertson, chief operations engineer for the Los Angeles Department of Water and Power [8]. More recently, the Japanese carried out the real development of the rubber dam concept. Rubber dams are mainly found in Japan, USA, Germany, France and in the former USSR. With several times the number of installations as in all other nations combined, Japan has been the most receptive to this water regulation concept [9]. Amongst the most experienced manufacturers of inflatable dams are Sumitomo and Bridgestone, both of Japan, with more than 1200 and 400 installations worldwide, respectively.

An inflatable gate consists basically of a rubberized fabric tube anchored to the sill of the dam. The fabric is secured to the foundation slab with clamping bars and anchor bolts either by double or single row anchoring.

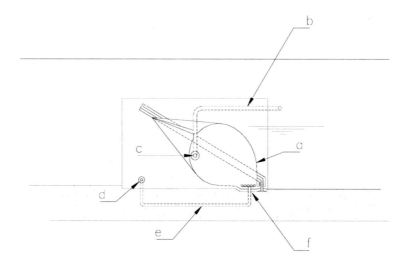

Fig. 15.16 Inflatable gate (SUMIGATE)
(a) rubber body; (b) air-supply-exhaust; (c) air-supply-exhaust opening in dam body; (d) drain pipe outlet; (e) drain pipe; (f) drain pipe intake

The gate ends are usually taken up the sloping sides of the channel and closed so that a complete seal is produced. Alternatively, side sealing may be provided by direct pressure of the rubber dam against the concrete sidewalls. Rubber dams can be installed without perforating the fabric as shown in Figure 15.17.

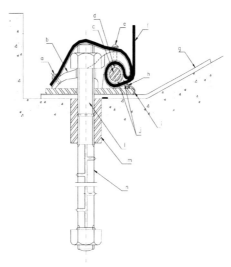

Fig. 15.17 Cross section of rubber dam anchoring (SUMITOMO)
(a) drive screw; (b) clamping bar; (c) rod; (d) tube; (e) drive screw; (f) carpet; (g) sealing sheet;
(h) bar; (i) cover sheet; (j) polyester resin; (l) bolt; (m) sleeve; (n) anchor bolt

A combination of nylon fabric and synthetic rubber, bonded together in a sandwich construction, is used for the inflatable weir body. Thickness and number of nylon layers of the fabric are selected in relation to the gate height. Reinforcing nylon provides the necessary strength for inflation and to withstand water pressure and the composite rubber layers are responsible for air tightness. Generally, the thickness of the fabric varies from 4 to 16mm.

In contrast to other types of weirs, the inflatable gate produces no concentrated loads but a uniformly applied load across the weir length. As a result, they are particularly well suited for installation on existing structures.

Inflation of the rubber gate is made with water, or more often, with air. Air inflated systems are usually cheaper and of simple maintenance. Operation equipment associated with air is smaller and simpler in design and operation than that for water. Water-filled weirs require space for pipes and filters necessary to remove sediment from water pumped into the inflatable gate. In cold environments, the use of air avoids the danger of water-filled pipe freezing. Also, air-filled weirs inflate and deflate quicker than those using use water. In some cases, inner inspection of air-filled gates is possible under an inflated condition.

Air pressure of rubber dams can be manipulated through a microprocessor installed in the control system to maintain a regulated upstream water level. In this case, the rubber dam deflates slightly to permit increased water passage over its body. If the water level begins to fall, the rubber dam is inflated to maintain the set point. Also, rubber dams can be equipped with mechanisms to ensure a complete and automatic deflation during flooding. Logs, tree stumps and other debris can be passed over the inflated rubber dam. To reduce environmental impacts, the rubber dam may be temporarily lowered to allow sediment to flush through the impoundment.

One of the most interesting advantages of the rubber dam is its speed of erection, since the time required for it varies from 5 to 15 days, not considering foundation, cofferdams, anchors and piping installation works. Maintenance costs, too, are fairly minimal, as the rubber dam needs no painting or rust protection and the only regular maintenance is to check the inflation/deflation equipment. According to manufacturers, the expected life of rubber dams is about 40 years.

An unusual construction of rubber dam was developed by Obermeyer Hydro Inc., USA, which consists of a series of steel panels supported on their downstream side by inflatable air bladders. By controlling the pressure in the bladders, the pond elevation maintained by the gates can be adjusted.

Inflatable gates are designed for heights from 0.5 m to 6 m and single spans up to 100 m. Outstanding examples of the use of inflatable dams are shown in Table 15.4.

Table 15.4 Examples of Use of Inflatable Weirs

Year	Project	Height x span (m x m)	Application	Manufacturer
1989	Alameda (USA)	3.9 x 88.8	Groundwater recharging	Bridgestone
1989	Rainbow (USA)	3.5 x 67.7	Hydroelectricity	Bridgestone
1984	Susquehanna River (USA)	2.4 x 88.7	Recreation	Bridgestone
1978	Tai Po Tau (Hong Kong)	3.7 x 38.1	Irrigation and flood control	Sumitomo
1986	Mirani (Australia)	1.8 x 107.2	Irrigation	Bridgestone
1989	Kjedal (Norway)	2.0 x 84.1	Navigation	Bridgestone
1981	Doufuku (Korea)	1.8 x 80.0	Increase of reservoir capacity	Sumitomo
1978	Mine (Japan)	2.9 x 30.3	Hydroelectricity	Sumitomo
1990	Tin Shui Wai (Hong Kong)	2.2 x 52.5	Pollution control	Bridgestone
1982	Gifu (Japan)	5.0 x 10.0	Hydroelectricity	Sumitomo
	Mulde (Germany)	2.0 x 50.0	Hydroelectricity	Floecksmühle
1989	Hiroshima (Japan)	1.0 x 65.5	Irrigation and flood control	Sumitomo

A new gate type developed in France in 1989 allows the increase of the storage capacity and the flooding discharge capacity of existing reservoirs. In its basic conception, the new gates (known as *Hydroplus* fusegates) are independent free-standing gravity units placed on the dam crest, and designed to operate like an ungated structure during moderate flows, and as a gated structure in case of large flows. When the headwater is lower than or level with the top of the fusegate, it acts as an extension of the dam.

Fusegates are installed side-by-side on a spillway sill to form a watertight barrier. They bear against small abutment blocks set on the sill to prevent them sliding under hydrostatic pressure. In the most common version, there is a small cavity under the base of each fusegate, with drain holes to discharge accidental inflow. An inlet stack or well on the upstream side of the fusegate crest discharges into the chamber when the headwater reaches a predetermined level. The lips of the stacks on individual fusegates can be set at different levels. Moderate floods spill over the fusegate barrier. Larger floods cause the chamber to fill as the water spills over the wells lips. The pressure builds up on the underside and this uplift pressure causes the fusegate to turn over its abutment frame. Each individual fusegate is designed with sufficient weight to remain stable until the chamber is filled, but then to become unstable when the uplift pressure reaches a predetermined level. When the design flood passes through the reservoir, the fusegates successively rotate. Once the flood has receded, the overturned unit (or units) can be reinstalled on the sill, or replaced, if damaged. Ballast can be added for precise control of the pressure required to overturn an individual fusegate.

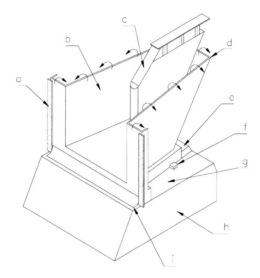

Fig. 15.18 Upstream view of crest fusegate (HYDROPLUS)
(a) side seal; (b) bucket; (c) inlet well; (d) overspilling crest; (e) downstream abutment block;
(f) side abutment block; (g) downstream bucket side; (h) concrete sill; (i) upstream seal strip

Fusegates are totally non-mechanical and do not require any human intervention nor electrical or mechanical power during operation. These gates can be manufactured either in steel or reinforced concrete. When made of reinforced concrete, maintenance is reduced to a minimum. This type of gate does not require any intermediate piers as all loads are transmitted directly to the spillway sill through the blocks located at the downstream edge.

Fusegates can also be arranged in a circle on morning-glory spillways. In 1991, twelve 1 m-high by 1 m-wide concrete fusegates were installed in the morning-glory spillway of the Gouyre dam, in France, allowing a storage increase of 17% and a spillway discharge capacity increase of 27%. They are specially designed to pass through the spillway shaft and culvert without causing any obstruction.

The fusegate stability against sliding is ensured simply by blocks located at the downstream edges, and concreted onto the spillway sill. The overturning moment are caused by:
. hydrostatic pressure on the upstream faces;
. uplift pressure in the chamber and under the fusegate base (when water enters the inlet well, the uplift pressure rises rapidly and causes the fusegate to overturn).
The following forces produce the resisting moment:
. dead weight of the fusegate;
. weight of ballast, if any;
. weight of the water in the bucket floor (applying a downward pressure on the fusegate); and,
. back-pressure from tailwater against the downstream face of the fusegate.

A. A. Alla made a detailed analysis of the fusegate stability [10]. According to the manufacturer, the fusegate height ranges from 1 to 8 m [11]. The first dam modified to incorporate the new system was the Lussas irrigation dam in Ardèche, France, in 1991, where a 35 m length of the sill was equipped with 10 fusegates, each gate being 2.15 m high and 3.5 m wide. In 1994, 10 fusegates 6.5 m-high by 9.73 m-wide were installed at the Shongweni dam, in South Africa, increasing its spillway discharge capacity from 1250 to 5000 m³/s.

Fig. 15.19 Fusegates at the Shongweni dam, South Africa (HYDROPLUS)

REFERENCES

1. van Ieperen, A.: Design of the New Waterway Storm Surge Barrier in The Netherlands, *The International Journal on Hydropower & Dams* (May 1994).
2. Hindley, M.: Swing Out Twister, *International Water Power and Dam Construction* (Sept. 1994).
3. Ministry of Transport, Public Works and Water Management, Netherlands, *The New Waterway Storm Surge Barrier* (Nov. 1995).
4. Krasil'nikov, M.F., Improvement of the Mechanical Equipment of High-Head Spillways, *Gidrotekhnicheskoe Stroitel'stvo*, No. 12 (Dec. 1982).
5. Buzzel, D.A.: Trends in Hydraulic Gate Design, *Transactions of ASCE*, Paper No. 2908 (1957).
6. Erbisti, P.C.F. and Baumotte, P.P., Heightening of Spillway Gates, *Conference on Uprating and Refurbishing Hydropower Plants* (1987), Strasbourg, France.
7. Leviel, C.: French Experience of Inflatable Dams, *The International Journal on Hydropower & Dams* (Sept. 1994).
8. Kahl, T. and Ruell, S.: Flashboard Alternatives including Rubber Dams, *Waterpower'89* (Aug. 1989), Niagara Falls, NY.
9. Takasaki, M.: The Omata Inflatable Weir at the Kawarabi Hydro Scheme, Japan, *International Water Power and Dam Construction* (Nov. 1989).
10. Alla, A.A.: The Role of Fusegates in Dam Safety, *The International Journal on Hydropower and Dams*, Issue Six (1996).
11. Lemperière, F.: Overspill Fusegates, *International Water Power and Dam Construction* (July 1992).

NAME INDEX

Ackermann, H., 7, 15
Alla, A.A., 342
ASHRAE, 244, 248
ASCE, 14

Barois, J., 15
Baumotte, P.P., 342
Berthold, R., 78
Biot, M. A., 152, 181
Blokland, P., 78
Boissonault, F.L., 94, 183, 197
Bureau of Reclamation, 15, 94
Buzzell, D.A., 81, 94, 342

Campbell, F.B., 238, 248
Cannell, P.J., 78
Cavalcante, A.F., 323
CIMAF, 273
Csallner, K., 15, 94
Cyranka, P., 323

Daumy, G., 81, 94
Davis, C.V., 94, 183, 197, 248
Davis, J., 77
Delaroche, J., 181
Djonin, K., 78
Douma, J., 78

Eberhardt, A., 78
Editora Abril Cultural, 14
Engels, H., 5, 15
Erbisti, P.C.F., 15, 78, 197, 342
Ethembabaoglu, S., 181
Eylers, C.F., 78

Faires, V.M., 273
Farmakovsky, S. A., 303
Fontaine, H., 28
Freishist, A.R., 78
Fujimoto, S., 248

GE, 289
Gomes Navarro, J.L., 15, 94,
    183, 197
Guyton, B., 238, 248

Hart, E.D., 78
Hartung, F., 15, 65, 78
Heckel, R., 181
Hetényi, M., 153, 181
Hindley, M., 342
Hom, O.S., 237, 248

van Ieperen, A., 342

Josserand, A., 181
Juan-Aracil, J., 15, 94, 183, 197

Kahl, T., 342
Kaplan, I.H., 303
Kent, 181
Knapp, F.H., 200, 219, 223
Kobus, H.E., 209, 223
Kollbrunner, C.F., 15, 94
Korotikov, B., 261, 273
Krasil'nikov, M.F., 342
Kulka, H., 5, 15

Lemperière, F., 342
Leohnhardt, F., 166, 181
Leviel, C., 342
Levin, L., 239, 243, 248
Liu, M., 187

Martenson, V.Ya., 78
de Mas, F.B., 14
Matos, G., 323
Monnig, E., 166, 181
Muskatirovic, J., 78
Murray, R.I., 223

Naudascher, E., 209, 223
Nelson, M., 78

O'Donnell, K.O., 78
Ortiz, C., 78
Oswalt, N.R., 78

Perez, P.M., 273
Pickering, G.A., 78
Predic, Z., 78

Rao, R.P., 209, 223
Rexroth Hydraudyne, 258, 270, 326
Roehle, W., 14
Rodney Hunt, 30, 78
Rudenko, N., 273
Ruell, S., 342

Sagar, B.T.A., 223
Sarkaria, G.S., 237, 248
Schleicher, F., 15
Schoklitisch, A., 15
Schreiber, G.P., 183, 197
Schuetz, K., 78
Schwarz, H.J., 78
Sharma, H.R., 150, 181, 240, 248
Simmons, W.P., 223
Smith, L.G., 15
Sorensen, K.E., 94, 183, 197, 248
Streiffer, A., 15
Streuli, L. J., 15

Takasaki, M., 342
Takasu, S., 248
Thomas, H.H., 94
Timoshenko, S.P., 181
Tullis, J.P., 223
Turazza, G., 14

U. S. Army Corps of Engineers, 73,
    208, 223, 248

Veltrop, J., 78
van de Ven, R., 273

Wegmann, E., 3, 14
Wiessing, J.M., 78

Young, W.C., 181

SUBJECT INDEX

Acceptance tests, 322-323
Aeration, 73, 235-248
Air demand ratio, 238-242
Air speed, 238-239, 246
Air vent
    characteristics, 236-237
    diameter, 237, 246
    dimensioning, 243-46
    empirical calculation, 237
    functions, 20, 236-237
Air viscosity, 244
Allowable deformation, 132
Allowable stresses, 113-115
Anchorages, 58-60, 81
André-Fricke, 153, 155-156
Angular deflection, 153, 171, 272
Anticorrosive protection, 305, 310-313
Arms
    arrangement, 51-54, 133-137
    axial load, 135-136, 143-145
    bracing, 52, 148
    buckling check, 146-148
    equally loaded, 135-136
    inclined, 53-54, 144, 226
    loads on, 53-54, 135, 143-144
    parallel, 53-54, 143-144
    position, 50-54, 143-144
Automatic operation, 40, 46-50

Beams
    bending, 131-132
    elastic curve, 171
    flange, 123-125
    quantity, 120-121
    spacing, 99-101
    vertical, 135-139
Bearings
    articulated, 327
    bronze bushing, 55
    concentric, 45
    eccentric, 331-332
    gudgeon, 36, 37
    loads, 56-57, 144-145
    pintle, 36-37
    roller, 56-57, 170-172, 250

spacing, 17, 60
    wood, 57
Belleville springs, 169
Bending moment, 131-132, 134-135,
    138-139, 146-148, 153-154,
    160-162
Bernoulli, 200
Bi-axial stresses, 113
Block-out, see Niche
Bolts
    material, 278
    spacing, 288
Bottom gate geometry, 205-206,
    208-213
Bottom heightening, 333, 335-336
Bottom outlet, 12, 28, 31-32, 68, 80,
    236, 299
Bracing, 148
Brake, 251, 255-256
Brass, 278, 284
Bronze, 278, 284
Buckling, 123-129, 146-148
Bulkhead, see Stoplogs
Buoyancy, 7, 23, 199, 225
Bushings
    allowable pressure, 180
    bronze, 55, 180, 226-227
    design, 180
    pressure, 180
    self-lubricating, 55, 81, 180
    wall thickness, 180
    wood, 57
By-pass valve, 26-27, 323

Cantilevered pin, 151, 172
Cast iron, 28-29, 278, 283
Cast steel, 28, 277, 282
Catapulting, 72-74, 202
Cavitation, 31, 149-150
Ceramic coated rod, 260, 270
Chain, 255-257
CIMS, 270
Clip, 253-254
Closure by gravity, 28, 33, 50, 68, 77,
    81, 250

Coefficient
    allowable stresses, 113-114
    buckling, 124-125, 127, 148
    determination, 183, 196
    downpull, 208-214
    form, 239
    friction, 226-228
    head losses, 243-246
    rolling friction, 227
    strength reduction, 168
    foundation, 153
Components, 10
Compressed flange, 123-125
Concrete
    bearing pressure, 155-156, 161,
        164-168
    characteristic compression strength,
        165, 167
    characteristic tensile strength, 166
    design compression strength, 167
    non-reinforced, 166-167
    pouring, 149
    reinforced, 167-168
Constant level, 11, 46-50
Contact pressure, 172-179
Counterguide, 10, 63, 68, 169
Counterweight, 4-5, 19, 42, 47-49, 62,
    271, 328-329
Cracking, 12, 27, 33, 72, 323
Crest heightening, 333, 337
Critical comparison stresses, 126, 128
Curvature radius, 42, 50-51, 60, 317
Curving, 305-306

Dam heightening, 332-342
Deflection, 116, 132, 152-154,
    160-162, 171
Degrees of surface preparation, 311
Design criteria, 268-269
Devaglide, 55
Dimensions, 51, 81-93, 122, 136-137,
    150-151
DIN 1045, 165
DIN 19704, 114, 165, 269
DIN 19705, 253
DIN 4114, 123-128, 131, 148
DIN 8570, 315
Discharge

capacity, 44, 80
    coefficients, 215
    floating debris, 21, 40, 61, 80
Dogging device, 27
Downpull
    coefficients, 208-214
    factors of influence, 205-207
    forces, see Hydrodynamic forces
    formulae for prediction, 208-222
Downstream seals, 68, 70-71, 74-75,
    189, 201, 301
Drum, 251-252

Eccentric hinge, 331-332
Eccentricity, 172
Effective width of skin plate, 119-120
Elastic
    continuum, 152
    foundation, 152-155, 160-162
    line, 171
    stability, 123-129
    support, 169
Electric motor, 251, 255-256, 260
Ellipse of contact, 174-175
Elliptical contact, 172-178
Embedded parts, 10, 40, 50, 67,
    149-181, 316-321
End girders, 11, 68
Equivalent thickness, 219-222
Erection
    embedded parts, 318-321
    instructions, 318
    sequence, 319
Euler reference stress, 126

Failure, 33, 72-73, 235-236
Field
    erection, 318-321
    tests, 322-323
Filters, 260-261, 264
Finishing, 307
Fittings, efficiency, 253
Flanges, 123-125
Floatation, 40
Float, 19, 42
Floating
    debris, 11, 21, 40, 61, 80
    gate, 64, 327, 331, 337

Flow
  over the gate, 18, 23, 45, 61
  stagnation, 149
  under the gate, 13, 44, 61, 69-71,
    80
Force
  measurement, 202-203
  seal friction, 225, 228-229
  support friction, 225-228
  wheel friction, 225, 227-228
Forged steel, 277, 282
Foundation
  coefficient, 153
  modulus, 152
  ratio of Poisson, 153
Frequency of movement, 178
Friction forces
  coefficient, 226-228, 243-244
  seal, 31, 34, 228-229
  ship, 111-112
  support, 55, 146, 225-228
  wheel, 227-228
Froude, 203, 238-241
Full-size laying out, 305-306
Functional tests, 314
Fusegate, 9, 340-342

Galle chain, 256
Gate
  application, 11, 79-81
  assembly, 321-322
  automatic, 42, 46-50
  bottom shape, 75, 205-206,
    208-214
  bear-trap, 8, 11, 65-67, 84, 90, 249
  bottom outlet, 12, 28, 31, 68, 80,
    235-236
  Broome, see Gate, caterpillar
  caterpillar, 9, 32-35, 83, 87,
    189-190, 196
  classification, 12-14
  constant level, 46-50
  cylinder, 9, 22-23
  double-leaf fixed wheel, 6, 9, 13,
    69-71, 79, 85, 92-93, 186-187,
    194
  double-leaf segment, 6, 9, 13-14,
    44-45

drum, 8, 11, 64-65, 84, 91, 93, 249
fixed-wheel, 9, 11, 67-76, 80, 88,
  131, 151, 185-187, 190,
  192-194, 300-302, 309
flap, 5, 17-20, 43, 70, 79, 84-85,
  90, 92-93, 108-109, 183-184,
  188-189, 195-196, 303
floating, 331, 337
framing, 39, 51, 133-139, 325
heightening, 333-336
high head, 14, 83, 235, 289,
  330-332
hoists, 249-273
inflatable, 9, 337-340
lock, 1-3, 12, 35-40, 45-46, 293
long-span, 40, 325-329
miter, 1-2, 5, 35-40, 293
mixed, 13
multiple, 13
radial, 14, 102-104, 327-329, 331
reverse segment, 3, 9, 14, 43-44,
  298, 335
ring, 9, 23
roller, 7, 13, 40-41, 83, 89
sector, 6-7, 9, 11, 60-62, 84, 91,
  249, 335
segment, 3 7, 9, 20-22, 42-60, 79,
  83, 86-87, 92, 102-104,
  132-139, 146-148, 184-185,
  191-192, 226-227, 236,
  265-267, 271-272, 296-299,
  330-337
slide, 12-13, 28-32, 80, 83, 86, 88,
  235, 325-326, 332
stainless steel, 28
Stoney, 3, 9, 13, 62-64, 83, 89
stoplogs, 12-13, 23-27, 151,
  160-162, 187-188, 194-195,
  294-295, 322-323
submersible segment, 45-46
submersible, 7, 41, 43-46
translation, 13
translo-rotation, 13
types, 12, 17-78, 79-85
vertical hinge, 5, 13, 36, 46
vertical lift, 95-101, 201-202
visor, 9, 13, 77-78, 329
wood, 1-5, 8, 28-29

Gear rack, 41
Girder
    arrangement, 99-101, 133-139
    depth, 122
    dimensions, 122-123
    flanges, 123-125
Gravity closure, 28, 32, 50, 68, 81, 250
Gray cast iron, 28, 278, 283
Grease fittings, 172
Gudgeon, 36-37
Guides, 10, 68, 149-181

Hand operation, 249-250, 268
Heat treatment, 275-276
Heightening, 332-337
Hertz, pressure, 173
Hinge axis, 5, 13, 19-20, 36, 42, 47-50,
    60, 64-65
History, 1-9
Hoist
    arrangement, 40, 265-268
    chains, 44, 81, 249, 255-257
    hydraulic, 31-32, 44, 81, 249,
        258-268, 270-275, 326
    screw lifts, 81, 249-251, 269-270
    types, 249
    wire ropes, 63, 77, 81, 249,
        251-254
Horizontal girders, 99-101, 131-135,
    146-147
Hydraulic,
    chamber, 23, 60, 64-65
    cylinder, 31-32, 44, 81, 249,
        258-268, 270-275, 326
    diagram, 262-264
    flow, 149-150, 240-242
    head losses, 243-246
    operation, 11, 23, 60, 64-67, 81
    power unit, 260-265
Hydrodynamic forces
    description, 112, 199-223
    factors of influence, 205-207
    formulae for prediction, 208-222
    measurement, 202-204
Hydroplus, see Fusegate
Hydrostatic load, 95-109

Ice, 40, 61, 80, 111, 113

Inertial angular detector, 272-273
Inflatable gate, 9, 337-340
Inspection, 313-317
Intake, 22, 68, 79
Irrigation, 12, 28, 46

Jet velocity, 200, 209-215, 219-220

Knapp's method, 219-223

Lateral guidance, 169
Leakage, 65, 291
Lifting beam, 24-26, 314, 317, 322
Limit switches, 251, 254-255, 260,
    270, 322-323
Limits of use, 79-85
Linear transducer, 272
Lintel beam, 11-12, 74
Load
    cases, 111-114, 269
    design, 268-269
    embedded parts, 79-80
Lock
    aqueducts, 5, 43, 80, 298
    gate, 1-3, 12-13, 80, 293
Lubrite, 55

Mach, number of, 246
Machining, 285, 305, 308-310
Manual operation, 249-250, 268
Manufacturing
    steps, 305-314
    tolerances, 315-317
Marking, 305-306
Mask beam, 333, 336
Materials, 260, 275-284, 289
Mechanical fit-up, 310
Mechanical element design, 114
Metallic wedge, 30
Method of Knapp, 219-223
Model tests, 73, 202-204, 208-209
Modernization, 332-342
Modulus of elasticity, 116, 126, 132,
    152, 155-157, 171, 173
Moody, diagram of, 243
Morning glory spillway, 9, 23, 341

Nappe, 7, 20, 50

NBR-8883, 111-120, 156-157, 165
Niche
    dimensions, 31, 150-151
    flow, 149-150
    geometry, 150-151
    offsetting, 150
    pressure, 150
Nondestructive tests, 313-314

Oil
    characteristics, 264
    minimum level, 260
    speed, 265
    pressure, 260
    viscosity, 264
Oiles, 55
Operating
    forces, 112, 225-234
    speed, 268-269
Operational tests, 322-323
Overflow, 13, 18, 20, 61, 65, 69-71,
    84-85
Overlapping effects, 155

Painting, 310-313
Pedestals, 249-251, 269-270
Piezometric head, 199, 207, 210
Pin supported at the ends, 151, 172
Pinion, pitch diameter, 257
Pintle, 35-37
Plate stresses, 115-116
Polyethylene seal, 28
Position
    indicator, 269-273
    arms, 51-54, 133-137
Power unit, see Hydraulic power unit
Preassembly, 305, 308
Precompression of rubber seal, 292
Preoperational tests, 322-323
Preparation of surfaces, 310-311
Pressure balance, 12, 26-27, 33, 72
Projection of top seal, 205-207
PTFE, 56, 226, 228, 289-290

Quantity of girders, 120-121

Radius, skin plate, 42, 50-51, 60, 317
Rail, see Wheel track

Raw material storing, 305-306
Rectangular contact, 172-174
Reduced comparison stresses, 126-128
Refurbishment, 332-342
Reinforced concrete, 164-165
Reynolds, number of, 243
Rigid
    connection, 134-135
    frame, 147
RIVERT, see inertial angular detector
Rod, ceramic coating, 260, 270
Rolled steel, 276, 279
Roller bearings, 57, 170, 171-172, 227
Roller trains, 32-33, 62-63
Roughness, 243-244
Rounded connections, 134-155
Rubber
    hardness, 290
    material, 290
    seals, 285-303

Safety factors, 129, 269
Scale of model, 203
Screw lifts, 249-251, 269-270
Seal
    clad, 289-290
    friction, 228 229
    hardness, 290
    joints, 291-292
    leakage, 65, 291
    seat, 10, 30-31
    between panels, 76
    material, 30-31, 289-290
    metallic, 19, 23, 29, 32, 65,
        285-286, 302-303
    polyethylene, 28
    precompression, 26, 292
    rubber, 19, 285-302
    shapes, 287
    specifications, 290-291
Seismic force, 113
Selection of type, 79-94
Self-lubricating bushing, 55, 81, 180
Sequence of erection, 318-322
Service life, 114
Shear
    allowable stress, 113, 122
    force, 153

stresses, 122, 155
Ship
    friction, 111-112
    impact, 111-112
Shoes, 169
Shop tests, 313-314
Sill, 10, 50-51
Skin plate
    description, 10
    effective width, 119-120
    radius, 42, 50-51, 60, 317
    shape, 14
    stress, 115-121
    thickness, 115, 120, 132-133
Slack wire ropes, 254
Slenderness ratio, 125, 148
Slide track, 10, 160-162
Slot
    dimensions, 150-151
    geometry, 150
    lining, 10
Specific mass, 226
Spherical bearing, 56, 327-328
Spillway, 12, 79, 85, 249, 322-323
Springs
    Belleville, 169
    precompression, 26
Stability, 123-129, 146-148
Stainless steel gate, 28
Standard of design, see DIN 19704 and
    NBR-8883
Standard of manufacture, see DIN
    19705
Steel
    mechanical parts, 276, 280
    cast, 28, 277, 282
    forged, 277, 282
    rolled, 260, 263
    stainless, 28, 277, 281
Stiffeners, 131
Stoplogs, 12-13, 23-27, 187-188,
    194-195, 294-295
Strength reduction, coefficient, 168
Stress
    allowable, 113-114
    bi-axial, 114
    comparison, 113-114
    critical comparison, 126, 128

Euler reference, 126
    ideal buckling, 126-127
    reduced comparison, 126, 128
Structural design, 111-148
Structural steel parts, 114
Support
    elastic, 169
    rounded, 134-135
Surface preparation, 311

Tainter gate, see Gate, segment
Tank, 258-262
Teflon, see PTFE
Temperature, 112
Tests
    functional, 314
    nondestructive, 313-314
    operational, 323
    preoperational, 322-323
    shop, 313-314
Thermal effects, 112
Timber, 29, 40
Tolerances
    erection, 319-321
    manufacture, 315-317
Top heightening, 333-335
Top seal projection, 205-207
Transportation, 24, 317-318, 321
Trends in gate design, 325-342
Trunnions, 5, 55-59, 80, 226-227
Types of gates, 12, 17-78, 79-85

Underflow, 7, 13, 61, 80, 83, 85
Upstream seals, 68, 71-75, 201, 300

Vertical beams, 133-139
Vibration, 7, 28, 80, 205
Viscosity
    air, 244
    oil, 264

Wall offset, 150, 207
Waves, 112
Web
    depth, 122-123
    stability, 125-129
    thickness, 122, 125-126
Weight estimation

    caterpillar gate, 189-190
    double fixed-wheel gate, 186-187
    embedded parts, 190
    fixed-wheel gate, 185-186
    flap gate, 188-189
    segment gate, 184-185
    stoplogs, 187-188
Welding, 307
Wheel
    design, 170-174
    diameter, 173
    flange, 169, 227
    hardness, 178-179, 227-228
    cantilevered pin, 151, 172
    end-supported pin, 172
    shaft, 67, 170-172
    radius, 170
    track, 10, 68, 70-71, 152-156,
        172-180, 227-228
    tread width, 173
Wind, 112
Wire rope , 62-63, 77, 251-254, 269
Wood bushing, 57
Wood, 57, 285

Lightning Source UK Ltd.
Milton Keynes UK
UKOW030737131011

180173UK00012B/4/P